# UNDERSTANDING VOLTAMMETRY:
## Problems and Solutions

# UNDERSTANDING VOLTAMMETRY:
## Problems and Solutions

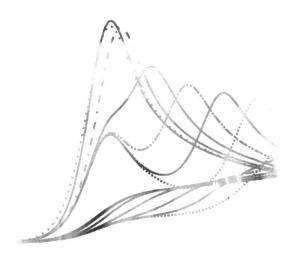

RICHARD G COMPTON
CHRISTOPHER BATCHELOR-MCAULEY
EDMUND J F DICKINSON

University of Oxford, UK

Imperial College Press

*Published by*

Imperial College Press
57 Shelton Street
Covent Garden
London WC2H 9HE

*Distributed by*

World Scientific Publishing Co. Pte. Ltd.
5 Toh Tuck Link, Singapore 596224
*USA office:* 27 Warren Street, Suite 401-402, Hackensack, NJ 07601
*UK office:* 57 Shelton Street, Covent Garden, London WC2H 9HE

**British Library Cataloguing-in-Publication Data**
A catalogue record for this book is available from the British Library.

UNDERSTANDING  VOLTAMMETRY
Problems  and  Solutions

ISBN-13 978-1-84816-730-8
ISBN-10 1-84816-730-X
ISBN-13 978-1-84816-731-5 (pbk)
ISBN-10 1-84816-731-8 (pbk)

Typeset by Stallion Press
Email: enquiries@stallionpress.com

Printed in Singapore.

# Publisher's Foreword

*Understanding Voltammetry: Problems and Solutions* is a companion volume to the textbook *Understanding Voltammetry 2nd Edition*, by Richard G. Compton and Craig E. Banks, published in 2011. The structure of this volume follows that of the textbook.

*Understanding Voltammetry* considers how to go about designing, explaining and interpreting experiments centred around various forms of voltammetry, including cyclic, microelectrode and hydrodynamic, amongst others.

The book gives clear introductions to the theories of electron transfer and of diffusion in its early chapters. These are developed to interpret voltammetric experiments at macroelectrodes before considering microelectrode behaviour. A subsequent chapter introduces convection and describes hydrodynamic electrodes. Later chapters describe the voltammetric measurement of homogeneous kinetics, the study of adsorption on electrodes and the use of voltammetry for electroanalysis.

# Contents

# Glossary of Symbols and Abbreviations

## Roman symbols

| | | |
|---|---|---|
| $A$ | area | $m^2$ |
| $A$ | Debye-Hückel constant $= 0.509\,\mathrm{mol}^{-\frac{1}{2}}\,\mathrm{kg}^{\frac{1}{2}}$ | |
| $a_i$ | activity of species $i$ | |
| $c_i$ | concentration of species $i$ | $\mathrm{mol\,dm^{-3}}$ |
| $c_{i,0}$ | surface concentration of species $i$ | $\mathrm{mol\,dm^{-3}}$ |
| $c^*$ | bulk concentration | $\mathrm{mol\,dm^{-3}}$ |
| $D$ | diffusion coefficient | $\mathrm{(c)m^2\,s^{-1}}$ |
| $E$ | cell potential | V |
| $E^\ominus$ | reduction potential under standard conditions | V |
| $E_f^\ominus$ | formal reduction potential | V |
| $F$ | the Faraday constant $= 96485.3\,\mathrm{C\,mol^{-1}}$ | |
| $G$ | Gibbs energy | J |
| $\Delta G^\ominus$ | change in Gibbs energy under standard conditions | $\mathrm{J\,mol^{-1}}$ |
| $\Delta G^\ddagger$ | activation energy | $\mathrm{J\,mol^{-1}}$ |
| $\Delta H^\ominus$ | change in enthalpy under standard conditions | $\mathrm{J\,mol^{-1}}$ |
| $h$ | height or half-height of a cell | m |
| $I$ | current passed | A |
| $I_{pf}$ | forward peak current | A |
| $I$ | ionic strength | $\mathrm{mol\,kg^{-1}}$ |
| $J$ | flux | $\mathrm{mol\,m^{-2}\,s^{-1}}$ |
| $K$ | equilibrium coefficient | |

| | | |
|---|---|---|
| $K$ | dimensionless rate constant | |
| $K_a$ | acid dissociation constant | |
| $K_{eq}$ | equilibrium coefficient (in follow-up kinetics) | |
| $K_{sp}$ | solubility product | |
| $k^0$ | heterogeneous rate constant | $(c)m\,s^{-1}$ |
| $k$ | rate constant | |
| $m_i$ | molality of species $i$ | $mol\,kg^{-1}$ |
| $n$ | number of electrons transferred | |
| $p$ | pressure as a multiple of standard pressure | bar |
| $pK_a$ | $\equiv -\log_{10} K_a$ | |
| $Q$ | reaction quotient | |
| $Q$ | charge transferred | C |
| $q_{rev}$ | reversible heat transferred | $J\,mol^{-1}$ |
| $q$ | charge | C |
| $R$ | the gas constant $= 8.31447\,J\,K^{-1}\,mol^{-1}$ | |
| Re | the Reynolds number | |
| $r$ | radius or radial coordinate | m |
| $r_e$ | electrode radius | m |
| $\Delta S^{\ominus}$ | change in entropy under standard conditions | $J\,K^{-1}\,mol^{-1}$ |
| $T$ | temperature | K |
| $t$ | time | s |
| $t_i$ | transport number of species $i$ | |
| $V_f$ | volume flow rate | $m^3\,s^{-1}$ |
| $v$ | voltammetric scan rate | $V\,s^{-1}$ |
| $W$ | rotation speed | $s^{-1}$ |
| $w$ | electrode width | m |
| $x$ | linear space coordinate | m |
| $z_i$ | charge number of species $i$ | |

## Greek symbols

| | | |
|---|---|---|
| $\alpha$ | Butler–Volmer transfer coefficient for reduction | |
| $\beta$ | Butler–Volmer transfer coefficient for oxidation | |
| $\Gamma$ | surface coverage | $mol\,(c)m^{-2}$ |
| $\gamma_i$ | activity coefficient of species $i$ | $m^3\,mol^{-1}$ |
| $\delta$ | Nernst diffusion layer thickness | m |
| $\epsilon_0$ | the permittivity of free space $= 8.854 \times 10^{-12}\,F\,m^{-1}$ | |
| $\epsilon_s$ | relative permittivity or dielectric constant of a solvent | |
| $\Lambda$ | the Matsuda–Ayabe parameter | |
| $\lambda$ | Marcus reorganisation energy | J |

| | | |
|---|---|---|
| $\mu_i$ | chemical potential of species $i$ | $J\,mol^{-1}$ |
| $\nu_i$ | stoichiometric coefficient of species $i$ | |
| $\nu$ | kinematic viscosity | $m^2\,s^{-1}$ |
| $\rho$ | density | $kg\,m^{-3}$ |
| $\tau$ | Shoup–Szabo time coordinate $\equiv (4Dt/r_e^2)$ | |
| $\phi$ | potential | V |
| $\phi_M$ | potential of a (metal) electrode | V |
| $\phi_s$ | potential of the solution phase | V |
| $\Delta\phi_{OD}$ | ohmic drop | V |
| $\Theta$ | fractional surface coverage | |
| $\Theta$ | dimensionless potential $\equiv \phi \times (F/RT)$ | |
| $\theta$ | dimensionless overpotential $\equiv (F/RT) \times (E - E_f^{\ominus})$ | |

## Abbreviations

| | |
|---|---|
| BDD | boron-doped diamond |
| BPPG | basal-plane pyrolytic graphite |
| EMF | electromotive force |
| EPPG | edge-plane pyrolytic graphite |
| erf($x$) | the error function |
| erfc($x$) | the complementary error function, $\equiv 1 - erf(x)$ |
| HOPG | highly ordered pyrolytic graphite |
| TBAP | tetra-$n$-butylammonium perchlorate |
| [$i$] | concentration of species $i$ |
| [$i$]$_0$ | surface concentration of species $i$ |
| $\ominus$ | standard state |

# 1

---

# Equilibrium Electrochemistry and the Nernst Equation

## 1.1 Cell Thermodynamics

### Problem

The measured electromotive force (EMF) for the cell

$$Pt_{(s)}|H_{2(g,p=1\,atm)}|H^+_{(aq,a=1)}||Cu^{2+}_{(aq,a=1)}|Cu_{(s)}$$

is $+0.337$ V. Write down the cell reaction and calculate the value of $\Delta G^\ominus$ for this reaction.

### Solution

The potential determining equilibria are as follows:
Right-hand electrode

$$\frac{1}{2}Cu^{2+}_{(aq)} + e^- \rightleftharpoons \frac{1}{2}Cu_{(s)} \tag{1.1}$$

Left-hand electrode

$$H^+_{(aq)} + e^- \rightleftharpoons \frac{1}{2}H_{2(g)}$$

Note these are written as reductions involving one electron. We therefore subtract to obtain a cell reaction:

$$\frac{1}{2}Cu^{2+}_{(aq)} + \frac{1}{2}H_{2(g)} \rightleftharpoons \frac{1}{2}Cu_{(s)} + H^+_{(aq)} \tag{1.2}$$

for which

$$\Delta G^\ominus = -FE^\ominus$$

All the species in Eq. 1.2 are present at unit activity, either as explicitly stated in the problem or implicitly in the case of copper since it is a pure solid and so must also be at unit activity. Note we assume that in the case of hydrogen at one atmosphere pressure that the gas will be sufficiently close to ideality that the effects of gas imperfections can be neglected.

It follows that

$$\Delta G^{\ominus} = -F \times 0.337 \text{ V}$$

$$= -32.5 \text{ kJ mol}^{-1}$$

which is favourable. We conclude that hydrogen gas can thermodynamically reduce aqueous Cu(II) to metallic copper and, consequently, that the metal will not dissolve in acid solutions to form $Cu^{2+}$.

## 1.2 The Nernst Equation

### Problem

For the following cell,

$$Cu_{(s)}|Cu^{2+}_{(aq)}||Ag^{+}_{(aq)}|Ag_{(s)}$$

at 298 K:

(i)   State the cell reaction.
(ii)  Give the Nernst equation for the cell.
(iii) Calculate the cell EMF when the ions are present at activities of
      (a) 1.0 and (b) 0.1.

The standard electrode potentials are:

$$E^{\ominus}_{Ag|Ag^+} = +0.80 \text{ V}$$

$$E^{\ominus}_{Cu|Cu^{2+}} = +0.34 \text{ V}$$

### Solution

(i)  The potential determining equilibria are:
     Right-hand electrode
$$Ag^{+}_{(aq)} + e^{-} \rightleftharpoons Ag_{(s)}$$
     Left-hand electrode
$$\frac{1}{2}Cu^{2+}_{(aq)} + e^{-} \rightleftharpoons \frac{1}{2}Cu_{(s)}$$
     Subtracting gives the cell reaction
$$Ag^{+}_{(aq)} + \frac{1}{2}Cu_{(s)} \rightleftharpoons Ag_{(s)} + \frac{1}{2}Cu^{2+}_{(aq)}$$

(ii) The Nernst equation for the cell is:

$$E = E^{\ominus} + \frac{RT}{F} \ln \left\{ \frac{a_{Ag^+}}{a_{Cu^{2+}}^{\frac{1}{2}}} \right\} \tag{1.3}$$

where

$$E^{\ominus} = E^{\ominus}_{Ag|Ag^+} - E^{\ominus}_{Cu|Cu^{2+}}$$

$$= 0.80\,V - 0.34\,V$$

$$= 0.46\,V$$

Note the absence of $Cu_{(s)}$ and $Ag_{(s)}$ from Eq. 1.3 since these are pure solids and hence have unit activity.

(iii) (a) Substituting

$$a_{Ag^+} = a_{Cu^{2+}} = 1$$

into Eq. 1.3 gives

$$E = 0.46\,V$$

(b) Similarly for

$$a_{Ag^+} = a_{Cu^{2+}} = 0.1$$

we find

$$E = 0.43\,V$$

## 1.3 The Nernst Equation

## Problem

For the following hypothetical cell,

$$Al_{(s)}|Al^{3+}_{(aq)}||Cu^{2+}_{(aq)}, Cu^+_{(aq)}|Pt_{(s)}$$

at 298 K:

(i) State the cell reaction.
(ii) Give the Nernst equation for the cell.
(iii) Calculate the cell EMF when
    (a) $a_{Al^{3+}} = a_{Cu^{2+}} = a_{Cu^+} = 1.0$
    (b) $a_{Al^{3+}} = a_{Cu^{2+}} = a_{Cu^+} = 0.1$

The standard electrode potentials are:

$$E^{\ominus}_{Cu^+|Cu^{2+}} = +0.15\,V$$

$$E^{\ominus}_{Al|Al^{3+}} = -1.61\,V$$

## Solution

(i) The potential determining equilibria are:

Right-hand electrode

$$Cu^{2+}_{(aq)} + e^- \rightleftharpoons Cu^+_{(aq)}$$

Left-hand electrode

$$\frac{1}{3}Al^{3+}_{(aq)} + e^- \rightleftharpoons \frac{1}{3}Al_{(s)}$$

Subtracting gives the cell reaction

$$Cu^{2+}_{(aq)} + \frac{1}{3}Al_{(s)} \rightleftharpoons Cu^+_{(aq)} + \frac{1}{3}Al^{3+}_{(aq)} \tag{1.4}$$

(ii) The Nernst equation is:

$$E = E^\ominus + \frac{RT}{F} \ln \left\{ \frac{a_{Cu^{2+}}}{a_{Cu^+} \cdot a_{Al^{3+}}^{\frac{1}{3}}} \right\} \tag{1.5}$$

where

$$E^\ominus = E^\ominus_{Cu^+|Cu^{2+}} - E^\ominus_{Al|Al^{3+}}$$

$$= 0.15 \text{ V} - (-1.61 \text{ V})$$

$$= 1.76 \text{ V}$$

Note that in Eq. 1.5 we have taken $a_{Al} = 1$. Further note that the activity $a_{Al^{3+}}$ is raised to the power ($\frac{1}{3}$), reflecting the stoichiometric coefficient of $Al^{3+}$ in Eq. 1.4.

(iii) (a) When all the solution phase species are present at unit activities:

$$E = E^\ominus = 1.76 \text{ V}$$

(b) When the activities of the three ions are all 0.1:

$$E = 1.76 \text{ V} + \frac{RT}{F} \ln \frac{0.1}{0.1 \times (0.1)^{\frac{1}{3}}}$$

$$= 1.76 \text{ V} + \frac{8.314 \times 298}{96485} \ln \frac{0.1}{0.1 \times (0.1)^{\frac{1}{3}}}$$

$$= 1.76 \text{ V} - 0.02 \text{ V}$$

$$= 1.74 \text{ V}$$

Note that although calculations can be made on this 'hypothetical' cell, the physical realisation of high concentrations of $Al^{3+}$ and $Cu^+$ in aqueous solution is unrealistic.

## 1.4 The Nernst Equation

### Problem

Draw a diagram of an electrochemical cell for which the Nernst equation is

$$E = E^\ominus - \frac{RT}{2F} \ln \frac{a_{Pb^{2+}}}{a_{Cd^{2+}}}$$

### Solution

A suitable cell is based on the following potential determining equilibria:
Right-hand electrode

$$\frac{1}{2} Cd^{2+}_{(aq)} + e^- \rightleftharpoons \frac{1}{2} Cd_{(s)}$$

Left-hand electrode

$$\frac{1}{2} Pb^{2+}_{(aq)} + e^- \rightleftharpoons \frac{1}{2} Pb_{(s)}$$

and the corresponding cell reaction is

$$\frac{1}{2} Cd^{2+}_{(aq)} + \frac{1}{2} Pb_{(s)} \rightleftharpoons \frac{1}{2} Cd_{(s)} + \frac{1}{2} Pb^{2+}_{(aq)}$$

The associated Nernst equation is

$$E = E^\ominus + \frac{RT}{F} \ln \frac{a^{\frac{1}{2}}_{Cd^{2+}}}{a^{\frac{1}{2}}_{Pb^{2+}}} \tag{1.6}$$

where $a_{Pb} = a_{Cd} = 1$ and so are omitted from Eq. 1.6. Rewriting Eq. 1.6 we see

$$E = E^\ominus - \frac{RT}{2F} \ln \frac{a_{Pb^{2+}}}{a_{Cd^{2+}}}$$

as required. The appropriate cell is therefore

$$Pb_{(s)} | Pb^{2+}_{(aq, a_{Pb^{2+}})} || Cd^{2+}_{(aq, a_{Cd^{2+}})} | Cd_{(s)}$$

## 1.5 Theory of the Nernst Equation

### Problem

(i)  How is electrical potential defined?
(ii) The chemical potential $\mu_i$ of a species $i$ in a non-ideal solution may be written:

$$\mu_i = \mu_i^0 + RT \ln a_i \tag{1.7}$$

where $R$ is the gas constant, $T$ is temperature, and $a_i$ is the activity of $i$, which is related to its concentration $c_i$ by the expression $a_i = \gamma_i c_i$, where $\gamma_i$ is the activity coefficient. Considering additionally the contribution of electrical potential to the Gibbs energy of $i$, give an expression for the electrochemical potential of a solution species.

(iii)  Considering the above, and the requirement of equality of chemical potentials at equilibrium, derive the Nernst equation.

(iv)  Why does the ratio of reactant and product activities in the Nernst equation take the form of an equilibrium coefficient?

(v)  The gas constant, $R$, appears in the Nernst equation. Why is the gas constant relevant to electrochemistry?

## Solution

(i)  Electrical potential may be defined as the electrical potential energy of a charge in an electric field, per unit charge. Electrical potential has units of volts (V), or equivalently joules per coulomb ($\mathrm{J\ C^{-1}}$). It is the energy required to bring a unit test charge from infinity to a particular point in space.

(ii)  Since the electrical potential is energy per charge, the energy may be given as potential, $\phi$, multiplied by charge; if we multiply by charge per mole ($z_i F$) we recover an electrochemical potential. Additionally, the logarithm may be expanded, so:

$$\mu_i = \mu_i^0 + RT \ln c_i + RT \ln \gamma_i + z_i F \phi$$

(iii)  Consider a one electron reaction

$$\sum_i v_i c_i + e^- \rightleftharpoons \sum_j v_j c_j$$

where $v_i$ is the stoichiometric coefficient of $i$.

Then at equilibrium

$$\sum_i v_i \mu_i + \mu_e = \sum_j v_j \mu_j$$

Writing the potential at the electrode as $\phi_M$ and that at the adjacent solution as $\phi_s$:

$$\sum_i v_i \mu_i^0 + RT \ln \left( \prod \gamma_i^{v_i} \right) + RT \ln \left( \prod c_i^{v_i} \right)$$

$$+ F\phi_s \sum_i v_i z_i + \mu_e^0 - F\phi_M$$

$$= \sum_j v_j \mu_j^0 + RT \ln \left( \prod \gamma_j^{v_j} \right) + RT \ln \left( \prod c_j^{v_j} \right)$$

$$+ F\phi_s \sum_j v_j z_j$$

and rearranging, noting that $\sum_j v_j z_j - \sum_i v_i z_i = -1$:

$$F(\phi_M - \phi_s) = \Delta\mu^0 + RT \ln\left(\frac{\prod \gamma_i^{v_i}}{\prod \gamma_j^{v_j}}\right)$$

$$+ RT \ln\left(\frac{\prod c_i^{v_i}}{\prod c_j^{v_j}}\right)$$

and

$$\phi_M - \phi_s = (\phi_M - \phi_s)^0 + \frac{RT}{F} \ln\left(\frac{\prod \gamma_i^{v_i}}{\prod \gamma_j^{v_j}}\right)$$

$$+ \frac{RT}{F} \ln\left(\frac{\prod c_i^{v_i}}{\prod c_j^{v_j}}\right)$$

Writing these potential differences as $E$ and recognising that the formal reduction potential $E_f^\ominus$ is that where all concentrations are unity, it follows that

$$E = E_f^\ominus + \frac{RT}{F} \ln\left(\frac{\prod c_i^{v_i}}{\prod c_j^{v_j}}\right) \tag{1.8}$$

(iv) An equilibrium coefficient naturally arises because the Nernst equation is a statement of equilibrium, but set up per unit charge for convenience in measurement. Multiplying through by $F$, we find:

$$FE^\ominus = RT \ln K$$

where $K$ is an equilibrium coefficient with its normal definition as a reaction quotient. Recognising that $FE^\ominus = -\Delta G^\ominus$ for a one-electron reduction, this is a conventional thermodynamic statement of equilibrium.

(v) The gas constant is a molar quantification of the contribution of entropy to an equilibrium. The entropy of a mixed solution is maximised by equalising the concentrations of different components as much as possible. So, unequal concentrations provide a driving force for a reaction to take place. This entropic effect contributes to the reduction potential as a term in the Nernst equation proportional to $R$.

## 1.6 The Debye–Hückel Limiting Law
### Problem

In Problem 1.5, the role of activity and activity coefficients in the Nernst equation was demonstrated. Most electrolytic solutions are non-ideal, such that $\gamma_i \neq 1$ for an ion. At low concentrations, the Debye–Hückel limiting law applies, which states that

$$\log_{10} \gamma_i = -A z_i^2 \sqrt{I}$$

where $A$ is a constant equal to 0.509 $\mathrm{mol}^{-\frac{1}{2}}\ \mathrm{kg}^{\frac{1}{2}}$ for aqueous solution, $z_i$ is the charge number of the ion, and $I$ is the ionic strength, defined

$$I = \frac{1}{2} \sum_j z_j^2 m_j$$

for all ions $j$ with molality $m_j$ $(\mathrm{mol\ kg}^{-1})$ in the solution.

(i)   What approximations are used in the derivation of the Debye–Hückel limiting law? Why is it inaccurate even at modest electrolyte concentrations?

(ii)  With reference to Eq. 1.7, show that ions are stabilised by increasing ionic strength according to the Debye–Hückel limiting law. How may this be explained physically?

(iii) Assuming the Debye–Hückel limiting law to hold for all ions, what is the difference in formal potential for the oxidation of ferrocene to ferrocenium if the supporting electrolyte $(\mathrm{TBA}^+\mathrm{ClO}_4^-)$ concentration is elevated from 5 mmol kg$^{-1}$ to 10 mmol kg$^{-1}$? You may ignore the influence of ferrocenium on the ionic strength. Explain the sign of the calculated change.

## Solution

(i)   The Debye–Hückel limiting law is derived by solving the Poisson–Boltzmann equation under certain simplifying assumptions. Specifically, ion–ion electrostatic interactions are considered the dominant cause of non-ideality, and ion–solvent interactions are ignored. The complete dissociation of the electrolyte is also assumed, and the ions are treated as point charges.

  Inaccuracies arise at modest electrolyte concentrations because the concentrations of ions predicted within an ionic atmosphere are unrealistic if a finite ionic radius is not considered. The assumption of a finite ionic radius leads to the extended Debye–Hückel limiting law. This is still inaccurate for concentrated ionic solutions because the depletion of solvent molecules due to solvation shells makes the assumption of zero ion-solvent interaction highly inaccurate. Only empirical formulas such as the Robinson–Stokes or Pitzer equations are able to address this issue. Most well-supported electrolyte solutions have behaviours in this latter regime.

(ii)  Since $A z_i^2 \sqrt{I}$ is necessarily positive, $\log_{10} \gamma_i < 0$. This implies that $a_i < c_i$. Since the contribution of activity to the electrochemical potential is $RT \ln a_i$ according to Eq. 1.7, it is clear that incorporation of non-ideality due to ion–ion interactions reduces electrochemical potential. This is equivalent to lowering energy and so the effect is stabilising.

This may be understood because the attractive force between oppositely charged ions causes these ions to 'encounter' one another in solution on average more often than the repulsive encounter between ions of like charge. Consequently, the time-averaged strength of attractive Coulombic forces in an ionic solution outweighs the strength of repulsive forces, leading to a net stabilisation of all ions present.

(iii) In Problem 1.5 it was shown that:

$$E_f^{\ominus} = (\phi_M - \phi_s)^0 + \frac{RT}{F} \ln\left(\frac{\prod \gamma_i^{\nu_i}}{\prod \gamma_j^{\nu_j}}\right)$$

where the species $i$ are being reduced to species $j$. For the case of the ferrocene/ferrocenium couple:

$$E_f^{\ominus} = (\phi_M - \phi_s)^0 + \frac{RT}{F} \ln\left(\frac{\gamma_{Fc^+}}{\gamma_{Fc}}\right)$$

Because ferrocene is uncharged, it is unaffected by non-ideality due to Coulombic forces, and so we can approximate $\gamma_{Fc} = 1$. Then:

$$E_f^{\ominus} = (\phi_M - \phi_s)^0 + \frac{RT}{F} \ln \gamma_{Fc^+}$$

Substituting in the Debye–Hückel equation and taking care with logarithm bases:

$$E_f^{\ominus} = (\phi_M - \phi_s)^0 + \frac{RT}{F} \ln 10 \log_{10} \gamma_{Fc^+}$$

$$= (\phi_M - \phi_s)^0 - \frac{RT}{F} \ln 10 \, A\sqrt{I}$$

For a change from $I_1$ to $I_2$:

$$\Delta E_f^{\ominus} = \frac{RT}{F} \ln 10 \, A \left(\sqrt{I_1} - \sqrt{I_2}\right)$$

and substituting in the values given, taking note that ionic strength and molality are equivalent for a monovalent binary electrolyte such as $TBAClO_4$:

$$\Delta E_f^{\ominus} = 0.0257 \times 2.303 \times 0.509 \times \left(\sqrt{0.005} - \sqrt{0.01}\right)$$

$$= -0.881 \, mV$$

Note that this is a very small change.

The change is negative because an elevation in ionic strength will stabilise the charged ferrocenium cation, but will not stabilise neutral ferrocene. Therefore, reduction of ferrocenium to ferrocene is less energetically favourable at higher ionic strength, so the formal reduction potential becomes more negative.

## 1.7  Cell Reaction and Equilibrium Constant

## Problem

At 298 K, the EMF of the cell shown below is +0.84 V.

$$Pt_{(s)}|Fe^{2+}_{(aq,a=1)}, Fe^{3+}_{(aq,a=1)}||Ce^{4+}_{(aq,a=1)}, Ce^{3+}_{(aq,a=1)}|Pt_{(s)}$$

(i) Define what is meant by the *standard* EMF of the cell.
(ii) Write down the cell reaction and the Nernst equation for the cell.
(iii) Calculate the equilibrium constant for the cell reaction at 298 K.

## Solution

(i) The 'standard' EMF of the cell is that measured when all of the chemical species in the cell are present at unit activity. Thus in the question $a_{Fe^{3+}} = a_{Fe^{2+}} = a_{Ce^{4+}} = a_{Ce^{3+}} = 1$, so that $E^{\ominus} = +0.84$ V.
(ii) The potential determining equilibria are as follows:
Right-hand electrode

$$Ce^{4+}_{(aq)} + e^- \rightleftharpoons Ce^{3+}_{(aq)}$$

Left-hand electrode

$$Fe^{3+}_{(aq)} + e^- \rightleftharpoons Fe^{2+}_{(aq)}$$

and so, subtracting, the reaction is

$$Ce^{4+}_{(aq)} + Fe^{2+}_{(aq)} \rightleftharpoons Ce^{3+}_{(aq)} + Fe^{3+}_{(aq)} \qquad (1.9)$$

The corresponding Nernst equation is

$$E = E^{\ominus} + \frac{RT}{F} \ln\left\{ \frac{a_{Ce^{4+}} \cdot a_{Fe^{2+}}}{a_{Ce^{3+}} \cdot a_{Fe^{3+}}} \right\}$$

$$= 0.84 \text{ V} + \frac{RT}{F} \ln\left\{ \frac{a_{Ce^{4+}} \cdot a_{Fe^{2+}}}{a_{Ce^{3+}} \cdot a_{Fe^{3+}}} \right\}$$

(iii) The equilibrium constant $K_c$ for Reaction 1.9 is related to $E^{\ominus}$ via

$$-\Delta G^{\ominus} = +FE^{\ominus} = +RT \ln K_c$$

$$\therefore \ln K_c = \frac{FE^{\ominus}}{RT}$$

$$= \frac{96485 \times 0.84}{8.314 \times 298}$$

$$K_c = 1.61 \times 10^{14}$$

## 1.8 Cell Reaction and Equilibrium Constant

### Problem

Consider the electrochemical cell

$$Zn|ZnSO_{4(aq, a_{Zn^{2+}})}||CuSO_{4(aq, a_{Cu^{2+}})}|Cu$$

for which the standard EMF is $E^\ominus = +1.10\,\text{V}$.

(i) Write down the cell reaction and the Nernst equation for the cell.
(ii) Calculate the standard Gibbs energy and the equilibrium constant for the cell reaction at 298 K.

### Solution

(i) The potential determining equilibria are as follows:
Right-hand electrode

$$\frac{1}{2}Cu^{2+}_{(aq)} + e^- \rightleftharpoons \frac{1}{2}Cu_{(s)}$$

Left-hand electrode

$$\frac{1}{2}Zn^{2+}_{(aq)} + e^- \rightleftharpoons \frac{1}{2}Zn_{(s)}$$

So from subtraction the cell reaction is

$$\frac{1}{2}Cu^{2+}_{(aq)} + \frac{1}{2}Zn_{(s)} \rightleftharpoons \frac{1}{2}Cu_{(s)} + \frac{1}{2}Zn^{2+}_{(aq)} \tag{1.10}$$

and the corresponding Nernst equation is

$$E = E^\ominus + \frac{RT}{F} \ln \left\{ \frac{a^{\frac{1}{2}}_{Cu^{2+}}}{a^{\frac{1}{2}}_{Zn^{2+}}} \right\}$$

$$= 1.10\,\text{V} + \frac{RT}{2F} \ln \left\{ \frac{a_{Cu^{2+}}}{a_{Zn^{2+}}} \right\}$$

(ii) $\Delta G^{\ominus}$ for Reaction 1.10 is given by

$$\Delta G^{\ominus} = -FE^{\ominus}$$

$$= -96485 \times 1.10 \text{ V}$$

$$= -106 \text{ kJ mol}^{-1}$$

The equilibrium constant $K_c$ for Reaction 1.10 is given by

$$\ln K_c = -\frac{\Delta G^{\ominus}}{RT} = +\frac{FE^{\ominus}}{RT}$$

$$= \frac{96485 \times 1.10}{8.314 \times 298}$$

$$K_c = 4.02 \times 10^{18}$$

## 1.9  Cell Reaction and Solubility Product

### Problem

From the following standard electrode potential data, calculate the solubility product ($K_{sp}$) of AgI at 298 K.

$$AgI_{(s)} + e^- \rightleftharpoons Ag_{(s)} + I^-_{(aq)} \quad E^{\ominus} = -0.152 \text{ V}$$

$$Ag^+_{(aq)} + e^- \rightleftharpoons Ag_{(s)} \quad \quad E^{\ominus} = +0.800 \text{ V}$$

### Solution

The standard electrode potentials are reported relative to the standard hydrogen electrode:

$$Pt_{(s)}|H_{2(g,p=1atm)}, H^+_{(aq,a=1)}||I^-_{(aq,a=1)}|AgI_{(s)}|Ag_{(s)}$$
$$E^{\ominus} = -0.152 \text{ V}$$
$$Pt_{(s)}|H_{2(g,p=1atm)}, H^+_{(aq,a=1)}||Ag^+_{(aq,a=1)}|Ag_{(s)}$$
$$E^{\ominus} = +0.800 \text{ V}$$

The cell reactions associated with these can be readily deduced to be

$$\frac{1}{2}H_{2(g)} + AgI_{(s)} \rightleftharpoons H^+_{(aq)} + I^-_{(aq)} + Ag_{(s)}$$

for which

$$\Delta G^{\ominus} = -FE^{\ominus} = -96485 \times (-0.152)$$
$$= +14.7 \text{ kJ mol}^{-1}$$

and

$$\frac{1}{2}H_{2(g)} + Ag^+_{(aq)} \rightleftharpoons H^+_{(aq)} + Ag_{(s)}$$
$$\Delta G^{\ominus} = -96485 \times 0.800$$
$$= -77.2 \text{ kJ mol}^{-1}$$

It follows that the reaction

$$AgI_{(s)} \rightleftharpoons Ag^+_{(aq)} + I^-_{(aq)}$$

can be written as the difference of the two reactions above, so that for the dissolution of AgI

$$\Delta G^{\ominus} = -RT \ln K_{sp}$$
$$= +14.7 - (-77.2)$$
$$= 91.9 \text{ kJ mol}^{-1}$$

Hence

$$K_{sp} \approx 7.8 \times 10^{-17}$$

## 1.10 Cell Reaction and p$K_a$

## Problem

The following electrochemical cell was studied at 298 K

$$Pt|H_{2(g,p=1atm)}|H^+_{(a=1)}||HA_{(a=1)}|H_{2(g,p=1atm)}|Pt$$

where HA is an organic acid.

If the acid dissociation constant ($K_a$) of HA is 0.27, determine the EMF of the cell.

## Solution

The potential determining equilibrium is

$$H^+_{(aq)} + e^- \rightleftharpoons \frac{1}{2}H_{2(g,p=1 \text{ atm})}$$

for both half-cells. If the activity of $H^+$ arising from the dissociation of HA

$$HA_{(aq)} \rightleftharpoons H^+_{(aq)} + A^-_{(aq)}$$

is $a_{H^+}$, then the cell reaction is

$$H^+_{(aq,a_{H^+})} \rightleftharpoons H^+_{(aq,a_{H^+}=1)}$$

where

$$\frac{a_{H^+} \cdot a_{A^-}}{a_{HA}} = K_a$$

Taking $a_{H^+} = a_{A^-}$ and $a_{HA} = 1$,

$$a_{H^+} = \sqrt{K_a}$$

The Nernst equation for the cell is:

$$E = E^\ominus + \frac{RT}{F} \ln\left\{ \frac{a_{H^+_{right}}}{a_{H^+_{left}}} \right\}$$

where $E^\ominus = 0$ V because both half-cells are hydrogen electrodes, and $a_{H^+_{left}} = 1$. Hence,

$$E = \frac{RT}{2F} \ln K_a$$
$$= -0.017 \text{ V}$$

## 1.11  Cell Thermodynamics and Temperature

## Problem

For the electrochemical cell

$$Pt|H_{2(g,p=1atm)}|HCl_{(aq,a=1)}|AgCl_{(s)}$$

the EMF at temperatures near 298 K obeys the following equation:

$$E^\ominus/V = -0.00558 + 2.6967 \times 10^{-3} T - 8.2299 \times 10^{-6} T^2$$
$$+ 5.869 \times 10^{-9} T^3$$

where $T$ is the absolute temperature measured in K. Calculate $\Delta G^\ominus$, $\Delta H^\ominus$ and $\Delta S^\ominus$ for the reaction at 298 K.

## Solution

The cell reaction is

$$AgCl + \frac{1}{2}H_{2(g,p=1)} \rightleftharpoons Ag_{(s)} + H^+_{(aq,a=1)} + Cl^-_{(aq,a=1)}$$

For this reaction

$$\Delta G^\ominus = -FE^\ominus$$

$$= -F(-0.00558 + (2.6967 \times 10^{-3} \times 298)$$

$$- (8.229 \times 10^{-6} \times 298^2) + (5.869 \times 10^{-9} \times 298^3))$$

$$= -F \times 0.223$$

$$= -21.5 \text{ kJ mol}^{-1}$$

Also

$$\Delta S^\ominus = -\frac{\partial \Delta G^\ominus}{\partial T} = F\frac{\partial E^\ominus}{\partial T}$$

$$= F(2.6967 \times 10^{-3} - (2 \times 8.2299 \times 10^{-6} T)$$

$$+ (3 \times 5.869 \times 10^{-9} T^2))$$

where the expression for $E^\ominus(T)$ has been explicitly differentiated with respect to temperature. At $T = 298$ K

$$\Delta S^\ominus = F(2.6967 \times 10^{-3} - (2 \times 8.2299 \times 10^{-6} \times 298)$$

$$+ (3 \times 5.869 \times 10^{-9} \times 298^2))$$

$$= -62.2 \text{ J K}^{-1} \text{ mol}^{-1}$$

The negative standard entropy change reflects the loss of gaseous hydrogen in the cell reaction. Finally,

$$\Delta H^\ominus = \Delta G^\ominus + T\Delta S^\ominus$$

$$= -21500 - (298 \times 62.2)$$

$$= -40 \text{ kJ mol}^{-1}$$

## 1.12 Cell Thermodynamics and Temperature

## Problem

The cell

$$Ag_{(s)}|Ag_2SO_{4(s)}|Hg_2SO_{4(aq,sat)}|Hg_2SO_{4(s)}|Hg_{(l)}|Pt_{(s)}$$

has an EMF of 0.140 V at 298 K. Close to 298 K, the EMF varies with temperature by $1.5 \times 10^{-4}$ V K$^{-1}$. Calculate $\Delta G^{\ominus}$, $\Delta S^{\ominus}$ and $\Delta H^{\ominus}$ for the cell reaction.

## Solution

The potential determining equilibria are as follows:
Right-hand electrode

$$\frac{1}{2}Hg_2SO_{4(s)} + e^- \rightleftharpoons Hg_{(l)} + \frac{1}{2}SO_{4(aq)}^{2-}$$

Left-hand electrode

$$\frac{1}{2}Ag_2SO_{4(s)} + e^- \rightleftharpoons Ag_{(s)} + \frac{1}{2}SO_{4(aq)}^{2-}$$

so that the cell reaction is

$$\frac{1}{2}Hg_2SO_{4(s)} + Ag_{(s)} \rightleftharpoons Hg_{(l)} + \frac{1}{2}AgSO_{4(s)}$$

where all the species are present in their standard states. It follows that

$$\Delta G^{\ominus} = -F \times 0.140 \text{ V}$$
$$= -13.5 \text{ kJ mol}^{-1}$$
$$\Delta S^{\ominus} = F\frac{\partial E^{\ominus}}{\partial T}$$
$$= F \times 1.5 \times 10^{-4}$$
$$= +14.47 \text{ J K}^{-1} \text{ mol}^{-1}$$
$$\Delta H^{\ominus} = \Delta G^{\ominus} + T\Delta S^{\ominus}$$
$$= -13500 + (298 \times 14.47)$$
$$= -9.19 \text{ kJ mol}^{-1}$$

## 1.13  Cell Energetics

### Problem

Consider the cell shown in Problem 1.12. How much heat is absorbed by the cell if it discharges isothermally and reversibly? How much heat ($q$) is absorbed if it discharges at constant external pressure, doing only reversible $pV$ work (that is, *no electrical work*)?

### Solution

If the cell operates reversibly and isothermally

$$q_{rev} = T\Delta S^{\ominus} = 4.31 \text{ kJ mol}^{-1}$$

If the cell operates doing no work but with constant external pressure, then

$$q = \Delta H^{\ominus}$$
$$= -9.19 \text{ kJ mol}^{-1}$$

## 1.14 Cell EMF and pH

### Problem

Consider the cell

$$Pt_{(s)}|H_{2(g)}|HCl_{(aq)}|AgCl_{(s)}|Ag_{(s)}$$

for which $E^{\ominus} = 0.2225$ V at 298 K. If the concentration of HCl is such that the measured cell potential is 0.385 V when the pressure of $H_2$ gas is one atmosphere, what is the pH of the solution?

### Solution

The cell reaction is

$$\frac{1}{2}H_{2(g)} + AgCl_{(s)} \rightleftharpoons H^+_{(aq)} + Cl^-_{(aq)} + Ag_{(s)}$$

and the associated Nernst equation is

$$E = E^{\ominus} + \frac{RT}{F} \ln \frac{p^{\frac{1}{2}}_{H_2}}{a_{H^+} \cdot a_{Cl^-}}$$

where $E^{\ominus} = 0.2225$ V at 298 K. Also, $p_{H_2} = 1$ and in the solution $[H^+] = [Cl^-]$. Since the ions have the same absolute charge we can assume that the activity coefficients are similar and so approximate $a_{H^+} = a_{Cl^-}$. It follows that

$$0.385 = 0.2225 - \frac{2RT}{F} \ln a_{H^+}$$

Hence

$$\ln a_{H^+} = -3.164$$

so that

$$pH = -\log_{10} a_{H^+}$$
$$= -\frac{\ln a_{H^+}}{\ln 10}$$
$$= 1.37$$

## 1.15  Cell Reaction and Equilibria

## Problem

Derive cells that can be used to

(i)  obtain the solubility product ($K_{sp}$) of $Cu(OH)_2$, and
(ii)  obtain the equilibrium constant ($K_c$) for the reaction

$$Zn_{(s)} + 4H^+_{(aq)} + PbO_{2(s)} \rightleftharpoons Pb^{2+}_{(aq)} + Zn^{2+}_{(aq)} + 2H_2O_{(l)}$$

## Solution

(i)  The required reaction is

$$\frac{1}{2}Cu(OH)_{2(s)} \rightleftharpoons \frac{1}{2}Cu^{2+}_{(aq)} + OH^-_{(aq)}$$

This can be written as the difference of two redox processes:

$$\frac{1}{2}Cu(OH)_{2(s)} + e^- \rightleftharpoons Cu_{(s)} + OH^-_{(aq)}$$

$$\frac{1}{2}Cu^{2+}_{(aq)} + e^- \rightleftharpoons \frac{1}{2}Cu_{(s)}$$

The required cell is

$$Cu_{(s)}|Cu^{2+}||OH^-_{(aq)}|Cu(OH)_{2(s)}|Cu_{(s)}$$

and if the EMF is measured under standard conditions

$$a_{Cu^{2+}} = a_{OH^-} = 1$$

then

$$\Delta G^{\ominus} = -FE^{\ominus} = -RT \ln K_{sp}$$

so that

$$K_{sp} = \exp\left(\frac{FE^{\ominus}}{RT}\right)$$

(ii)  The net reaction can be written as the difference of the redox reactions

$$\frac{1}{2}PbO_{2(s)} + 2H^+_{(aq)} + e^- \rightleftharpoons \frac{1}{2}Pb^{2+}_{(aq)} + H_2O_{(l)}$$

$$\frac{1}{2}Zn^{2+}_{(aq)} + e^- \rightleftharpoons \frac{1}{2}Zn_{(s)}$$

so that the required cell is

$$Zn_{(s)}|Zn(NO_3)_{2(aq)}||HNO_{3(aq)}, Pb(NO_3)_{2(aq)}|PbO_{2(s)}|Pb_{(s)}$$

for which the cell reaction is

$$\frac{1}{2}PbO_{2(s)} + 2H^+_{(aq)} + \frac{1}{2}Zn_{(s)} \rightleftharpoons \frac{1}{2}Pb^{2+}_{(aq)} + \frac{1}{2}Zn^{2+}_{(aq)} + H_2O_{(l)}$$

so that if standard conditions (unit activity for $Zn^{2+}$ and $Pb^{2+}$) are used in the cell, the measured EMF is

$$E^\ominus = \frac{RT}{F} \ln K'_c$$

where $K'_c$ is the equilibrium constant for the cell reaction as written above. Note that the stoichiometry for this reaction as given in the problem differs by a factor of 2 so

$$K_c = (K'_c)^2 = \exp\left(\frac{2FE^\ominus}{RT}\right)$$

## 1.16  Cell Reaction and $K_w$

### Problem

For the reduction of $ClO_4^-$ to $ClO_3^-$ the standard electrode potential under alkaline conditions is $+0.37$ V, while under acidic conditions it is $+1.20$ V. Write a balanced half-cell reaction for each reduction and deduce the value for the ionic product of water, $K_w$.

### Solution

Under alkaline conditions

$$\frac{1}{2}ClO_4^- + \frac{1}{2}H_2O + e^- \rightleftharpoons \frac{1}{2}ClO_3^- + OH^-$$

whereas in acid

$$\frac{1}{2}ClO_4^- + H^+ + e^- \rightleftharpoons \frac{1}{2}ClO_3^- + \frac{1}{2}H_2O$$

It follows that the reaction

$$H_2O \rightleftharpoons H^+ + OH^-$$

can be found by subtracting the latter reaction from the former, and that for the dissociation of water, as above:

$$\Delta G^\ominus = -F(0.37 - 1.20) = 80 \text{ kJ mol}^{-1}$$
$$= -RT \ln K_w$$

so

$$K_w = a_{H^+} \cdot a_{OH^-}$$

$$= \exp\left(\frac{-80000}{8.314 \times 298}\right)$$

$$\simeq 10^{-14}$$

## 1.17 Cell Reaction and Disproportionation

### Problem

Calculate the equilibrium constant for the disproportionation ($K_{disp}$) of MnOOH to $Mn^{2+}$ and $MnO_2$ at pH 3, given:

$$\frac{1}{2}MnO_{2(s)} + 2H^+_{(aq)} + e^- \rightleftharpoons \frac{1}{2}Mn^{2+}_{(aq)} + H_2O_{(l)} \qquad E_1^\ominus = 1.23 \text{ V} \qquad (1.11)$$

$$MnOOH_{(s)} + 3H^+_{(aq)} + e^- \rightleftharpoons Mn^{2+}_{(aq)} + 2H_2O_{(l)} \qquad E_2^\ominus = 1.5 \text{ V} \qquad (1.12)$$

### Solution

First, we write a balanced equation for the required equilibrium:

$$MnOOH_{(s)} + H^+_{(aq)} \overset{K_{disp}}{\rightleftharpoons} \frac{1}{2}Mn^{2+}_{(aq)} + \frac{1}{2}MnO_{2(s)} + H_2O_{(l)}$$

This equation is the difference of Eqs. 1.11 and 1.12. So:

$$\Delta G^\ominus_{disp} = (\Delta G^\ominus_2 - \Delta G^\ominus_1)$$

$$= 2F(E^\ominus_{1,pH=3} - E^\ominus_{2,pH=3})$$

We know from the Nernst equation (see also Problem 1.26) that:

$$\frac{\partial E}{\partial \text{pH}} = -\frac{mRT}{F} \ln 10$$

where $m$ is the number of protons transferred. Hence

$$E_{1,pH=3} = 1.23 - \frac{2RT}{F} \ln 10 \times 3$$

$$= 0.875 \text{ V}$$

and similarly

$$E_{2,pH=3} = 1.5 - \frac{3RT}{F} \ln 10 \times 3$$

$$= 0.968 \text{ V}$$

Hence

$$\Delta G^{\ominus}_{disp} = F(0.875 - 0.968)$$

$$= -8.98 \text{ kJ mol}^{-1}$$

From thermodynamics we know that

$$\Delta G^{\ominus} = -RT \ln K$$

so

$$K_{disp} = \frac{[Mn^{2+}][MnO_2][H_2O]^2}{[MnOOH]^2[H^+]^2} = 36.8$$

## 1.18 Fuel Cell Energetics

### Problem

In a fuel cell, hydrazine, $N_2H_4$, is oxidised to nitrogen, and oxygen is reduced to water. The standard electrode potentials for the reduction of $N_2$ to $N_2H_4$ and of $O_2$ to $H_2O$ at 298 K are $-1.155$ V and $+0.401$ V, respectively, both under alkaline conditions.

(i) Write a balanced equation for both of the half cell reactions under alkaline conditions. For the cell reaction where the hydrazine electrode is on the left, calculate the standard EMF of the cell at 298 K.

(ii) In a practical cell the concentrations of $N_2H_4$ and $OH^-$ are 0.5 M and 1.0 M, respectively, and the pressures of $O_2$ and $N_2$ are 0.2 bar and 0.8 bar, respectively. Use the Nernst equation to estimate the cell EMF at 298 K, assuming all activity coefficients are unity.

### Solution

(i) The half-cell reactions are

$$\frac{1}{4}N_{2(g)} + H_2O_{(l)} + e^- \rightleftharpoons \frac{1}{4}N_2H_{4(aq)} + OH^-_{(aq)}$$

and

$$\frac{1}{4}O_{2(g)} + \frac{1}{2}H_2O_{(l)} + e^- \rightleftharpoons OH^-_{(aq)}$$

The cell reaction is therefore

$$\frac{1}{4}N_2H_{4(aq)} + \frac{1}{4}O_{2(g)} \rightleftharpoons \frac{1}{2}H_2O_{(l)} + \frac{1}{4}N_{2(g)}$$

The Nernst equation for the cell is

$$E = E^{\ominus} + \frac{RT}{F} \ln \left\{ \frac{[N_2H_4]^{\frac{1}{4}} \cdot p_{O_2}^{\frac{1}{4}}}{a_{H_2O} \cdot p_{N_2}^{\frac{1}{4}}} \right\}$$

where

$$E^{\ominus} = 0.401 - (-1.155)$$

$$= +1.556 \, V$$

(ii) For the concentrations and pressures specified

$$E = 1.556 + \frac{RT}{4F} \ln \left\{ \frac{0.5 \times 0.2}{0.8} \right\}$$

$$= 1.543 \, V$$

## 1.19 Fuel Cell Energetics

### Problem

The overall cell reaction occurring in a 'direct methanol' fuel cell is

$$\frac{1}{6}CH_3OH + \frac{1}{4}O_2 \rightarrow \frac{1}{3}H_2O + \frac{1}{6}CO_2$$

$$\Delta G^{\ominus} = -117 \, kJ \, mol^{-1}$$

$$\Delta H^{\ominus} = -121 \, kJ \, mol^{-1}$$

(i) Write an equation describing how the cell potential varies with temperature. What is the cell potential for this reaction at 100°C?

(ii) What drawbacks arise from operating a fuel cell at low temperatures?

### Solution

(i) The two half-cell reactions are

$$\frac{1}{4}O_2 + e^- + H^+ \rightleftharpoons \frac{1}{2}H_2O$$

$$\frac{1}{6}CO_2 + e^- + H^+ \rightleftharpoons \frac{1}{6}CH_3OH + \frac{1}{6}H_2O$$

For this question we must assume that $\Delta S^{\ominus}$ and $\Delta H^{\ominus}$ are independent of temperature. From thermodynamics we know

$$\Delta G^{\ominus} = -FE^{\ominus} \qquad (1.13)$$

$$\left.\frac{\partial \Delta G^{\ominus}}{\partial T}\right|_p = -\Delta S^{\ominus} \qquad (1.14)$$

Substituting Eq. 1.13 into Eq. 1.14 we get

$$\left.\frac{\partial E^{\ominus}}{\partial T}\right|_p = \frac{\Delta S^{\ominus}}{F}$$

Integration gives

$$E(T) = E^{\ominus} + \frac{\Delta S^{\ominus}}{F}(T - T^{\ominus}) \qquad (1.15)$$

where $T^{\ominus} = 25°C$.
At standard conditions the cell potential is

$$E^{\ominus} = \frac{\Delta G^{\ominus}}{-F}$$

$$E^{\ominus} = \frac{-117000}{-96485} = 1.21 \text{ V}$$

and

$$\Delta S^{\ominus} = \frac{\Delta H^{\ominus} - \Delta G^{\ominus}}{T}$$

$$= \frac{-4000}{298} = -13.4 \text{ J K}^{-1} \text{ mol}^{-1}$$

Hence through the use of Eq. 1.15 we get, at $100°C$.

$$E^{\ominus}_{100} = 1.20 \text{ V}$$

(ii) Both methanol oxidation and oxygen reduction exhibit slow kinetics. Consequently, operation of the cell at low temperatures leads to lower current densities. Overcoming this requires either using large amounts of platinum or the development of more active catalysts.

## 1.20 The Influence of Temperature on the Self-Ionisation of Water

## Problem

(i) What pH would you expect a solution containing 0.1 M HCl to be? In your answer define pH and explain what an activity coefficient is.

(ii) The $pK_a$ of water is 15.66 at 298 K and the standard entropy of ionisation $(\Delta S^{\ominus})$ is $\approx -80$ J K$^{-1}$ mol$^{-1}$. How would you expect the pH of pure water to vary with temperature?

## Solution

(i)

$$pH = -\log a_{H^+}$$

$$a_{H^+} = \gamma_{\pm}[H^+]$$

The activity coefficient ($\gamma_{\pm}$) may be viewed as a measure of how far the solution deviates from ideality. This value is often approximated as being unity (i.e. the solution is assumed to be ideal). In 0.1 M HCl, the activity coefficient of a hydronium ion is actually 0.76 [H.S. Harned, *J. Am. Chem. Soc.* **38** (1916) 1986]; consequently, a solution of 0.1 M HCl will have a pH of 1.1, rather than exactly 1.

(ii)

$$H_2O \rightleftharpoons OH^- + H^+$$

$$K_{eq} = \frac{[OH^-][H^+]}{[H_2O]}$$

The van't Hoff equation as given below describes how the equilibrium constant varies with temperature.

$$\frac{\partial \ln K_{eq}}{\partial T} = \frac{\Delta H^\ominus}{RT^2}$$

The $pK_a$ of water is 15.66

$$\Delta G^\ominus = -RT \ln K_{eq}$$

and so $\Delta G^\ominus$ is positive. Hence from the equation below and given the magnitude of $\Delta S^\ominus$, $\Delta H^\ominus$ is positive.

$$\Delta G^\ominus = \Delta H^\ominus - T\Delta S^\ominus$$

Consequently, it can be seen that increasing the temperature will increase the value of $K_{eq}$. An increase in $K_{eq}$ leads to an increase in the proton concentration, and hence as we increase the temperature of pure water, its pH will decrease.

## 1.21  Cell Reaction and Complexation

## Problem

Calculate the standard electrode potential for the aqueous couple

$$[Fe(ox)_3]^{3-}_{(aq)}/[Fe(ox)_3]^{4-}_{(aq)}$$

from the following data (298 K), where $ox^{2-}$ refers to the oxalate anion, $C_2O_4^{2-}$:

$$Fe^{3+}_{(aq)} + e^- \rightleftharpoons Fe^{2+}_{(aq)} \qquad E^\ominus = +0.770 \text{ V}$$

$$Fe^{2+}_{(aq)} + 3ox^{2-} \rightleftharpoons [Fe(ox)_3]^{4-}_{(aq)} \qquad K = 1.7 \times 10^5$$

$$Fe^{3+}_{(aq)} + 3ox^{2-} \rightleftharpoons [Fe(ox)_3]^{3-}_{(aq)} \qquad K = 2.0 \times 10^{20}$$

## Solution

The standard electrode potential required can be found from $\Delta G^\ominus$ for the reaction

$$[Fe(ox)_3]^{3-}_{(aq)} + \frac{1}{2}H_{2(g)} \rightleftharpoons [Fe(ox)_3]^{4-}_{(aq)} + H^+_{(aq)}$$

From the data given

$$Fe^{3+}_{(aq)} + \frac{1}{2}H_{2(g)} \rightleftharpoons Fe^{2+}_{(aq)} + H^+_{(aq)}$$
$$\Delta G_1^\ominus = -0.770F$$
$$= -74.3 \text{ kJ mol}^{-1}$$

$$Fe^{2+}_{(aq)} + 3ox^{2-} \rightleftharpoons [Fe(ox)_3]^{4-}_{(aq)}$$
$$\Delta G_2^\ominus = -RT \ln K$$
$$= -29.8 \text{ kJ mol}^{-1}$$

$$Fe^{3+}_{(aq)} + 3ox^{2-} \rightleftharpoons [Fe(ox)_3]^{3-}_{(aq)}$$
$$\Delta G_3^\ominus = -RT \ln K$$
$$= -115.8 \text{ kJ mol}^{-1}$$

Then

$$\Delta G^\ominus = \Delta G_1^\ominus + \Delta G_2^\ominus - \Delta G_3^\ominus$$
$$= -74.3 - 29.8 - (-115.8) \text{ kJ mol}^{-1}$$
$$= +11.7 \text{ kJ mol}^{-1}$$

Hence the sought standard potential is

$$E^\ominus = -\frac{\Delta G^\ominus}{F}$$
$$= \frac{-11700}{96485}$$
$$= -0.121 \text{ V}$$

## 1.22 Reference Electrodes

### Problem

(i) The calomel and silver/silver chloride electrodes are commonly used in aqueous solution voltammetry. Identify the potential determining equilibria and the associated Nernst equation for each electrode. Comment on any implications of the latter.

(ii) Suggest a reference electrode suitable for voltammetry in non-aqueous solutions. Can the calomel and Ag/AgCl redox couples be used in these media?

### Solution

(i) For the calomel electrode

$$\frac{1}{2} Hg_2Cl_{2(s)} + e^- \rightleftharpoons Hg_{(l)} + Cl^-_{(aq)}$$

and

$$E_{Hg|Cl^-|Hg_2Cl_2} = E^\ominus - \frac{RT}{F} \ln a_{Cl^-}$$

Similarly for the Ag/AgCl electrode

$$AgCl_{(s)} + e^- \rightleftharpoons Ag_{(s)} + Cl^-_{(aq)}$$

and

$$E_{Ag|Cl^-|AgCl} = E^\ominus - \frac{RT}{F} \ln a_{Cl^-}$$

In both cases the reference electrode potential depends on the activity (concentration) of chloride ions present. It follows that a fixed (and known) concentration of the latter must be present in the reference electrode. Often solid KCl is added to the solution present in the reference electrode so as to ensure a saturated solution of KCl (see Fig. 1.1). Note however that the potential of such a *saturated* calomel (or Ag/AgCl) electrode will differ from a *standard* electrode, where $a_{Cl^-} = 1$.

(ii) A suitable reference electrode in non-aqueous solution, which is commonly employed in acetonitrile, is

$$Ag^+_{(CH_3CN)} + e^- \rightleftharpoons Ag_{(s)}$$

for which

$$E_{Ag/Ag^+} = E^\ominus + \frac{RT}{F} \ln a_{Ag^+}$$

Again the concentration of $Ag^+$ in the reference electrode must be known and fixed. Typically, 0.01 M $AgNO_3$ is used.

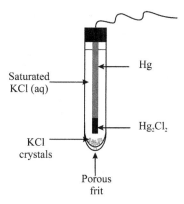

**Fig. 1.1** A saturated calomel reference electrode (SCE). Reproduced from R.G. Compton *et al.*, *Understanding Voltammetry*, 2nd ed., with permission from Imperial College Press.

Note that the Ag/AgCl reference couple is unsuitable in most non-aqueous, aprotic solvents since AgCl has a high solubility in these media, partly as a result of complex formation such as

$$AgCl + Cl^- \rightleftharpoons AgCl_2^-$$

Similarly the calomel reference couple is not used in organic solvents since $Hg_2Cl_2$ tends to disproportionate (forming Hg and Hg(II)) in these solvents.

## 1.23 Formal Potentials

### Problem

The standard electrode potential of $Ce(H_2O)_6^{4+}/Ce(H_2O)_6^{3+}$ couple is $+1.72$ V. The following formal potentials for the Ce(IV)/Ce(III) couple have been measured in different media: $+1.28$ V (1 M HCl), $+1.44$ V (1 M $H_2SO_4$), $+1.60$ V (1 M $HNO_3$), $+1.70$ V (1 M $HClO_4$). Comment.

### Solution

The standard potential quoted relates to the potential determining equilibrium

$$Ce(H_2O)_{6(aq)}^{4+} + e^- \rightleftharpoons Ce(H_2O)_{6(aq)}^{3+}$$

The implication of the term 'standard' is that the value relates to the condition of unit proton activity. Although protons do not feature in the above equilibrium, the Ce(IV) and Ce(III) species are prone to hydrolytic equilibria such as

$$M(H_2O)_{6(aq)}^{n+} \rightleftharpoons M(H_2O)_5(OH)_{(aq)}^{(n-1)+} + H_{(aq)}^+$$

as well as polymerisation and coordination with other anions present in solution. It is thought that the $Ce(H_2O)_6^{4+}$ ion only exists as such in concentrated perchloric acid solution. Note that the formal potential in this medium is close to the standard electrode potential. In solutions of other acids there is coordination of anions. This accounts for the dependence of the formal potential on the composition of the acid solution. It is noteworthy that since the standard potential of the reaction

$$\frac{1}{4}O_{2(g)} + H^+_{(aq)} + e^- \rightleftharpoons \frac{1}{2}H_2O_{(l)}$$

is $E^\ominus = +1.23$ V, solutions of Ce(IV) are capable of (slowly) oxidising water; such solutions are unstable and, although used widely in 'wet' analysis, need regular replacement.

## 1.24 Formal Potentials

### Problem

The standard electrode potential of the $Fe(H_2O)_6^{3+}/Fe(H_2O)_6^{2+}$ couple is $+0.77$ V. The following formal potentials for the Fe(III)/Fe(II) couple have been measured in different media: $+0.46$ V (2 M $H_3PO_4$), $+0.68$ V (1 M $H_2SO_4$), $+0.71$ V (0.5 M HCl), $+0.73$ V (1 M $HClO_4$), $+0.01$ V (1 M $K_2C_2O_4$, pH 5). Comment.

### Solution

$Fe^{3+}$ in solution tends to hydrolyse, polymerise and to form complexes:

$$Fe(H_2O)_6^{3+} \rightleftharpoons [Fe(H_2O)_5OH^-]^{2+} + H^+$$
$$K \simeq 9 \times 10^{-4}$$

$$[Fe(H_2O)_5OH^-]^{2+} \rightleftharpoons [Fe(H_2O)_4(OH^-)_2]^+ + H^+$$
$$K \simeq 5 \times 10^{-4}$$

$$2Fe(H_2O)_6^{2+} \rightleftharpoons [Fe(H_2O)_4(OH^-)_2Fe(H_2O)_4]^{2+} + 2H^+$$
$$K \simeq 10^{-3}$$

In $HClO_4$, HCl and $H_2SO_4$ media the pH is sufficiently low to prevent most (but not all) of these reactions and so the formal potential is not too far removed from the standard electrode potential.

At higher pH (1M $K_2C_2O_4$, pH 5) hydrolysis and polymerisation will occur and this is reflected in the huge discrepancy between the standard and formal potentials. In addition, both $Fe^{2+}$ and $Fe^{3+}$ can form couples with the oxalate anion (see Problem 1.17) and these effects also contribute to the large observed difference. In $H_3PO_4$ there is likely complexation of both $Fe^{2+}$ and $Fe^{3+}$ with the anions.

## 1.25 Standard Potentials and pH

### Problem

The 'quinhydrone electrode' is sometimes used to measure pH. Quinhydrone is an easily prepared, slightly soluble, equimolar (one-to-one) mixture of benzoquinone, $C_6H_4O_2$, and hydroquinone, $C_6H_4(OH)_2$. The reduction of the quinone shows fast ('reversible') electrode kinetics.

$$\frac{1}{2}C_6H_4O_2 + H^+ + e^- \rightleftharpoons \frac{1}{2}C_6H_4(OH)_2$$

$$E^\ominus = 0.700 \text{ V}$$

In practice the potential is reported with respect to a saturated calomel electrode (SCE).

(i) Show that the potential of the quinhydrone electrode gives the pH directly.
(ii) Suppose that a cell potential of $+0.160$ V is measured (vs SCE) for an unknown solution. What is the pH of the solution given that the $E^\ominus$ for the SCE is $+0.240$ V?
(iii) The quinhydrone electrode is not usable above pH 9. Speculate why.

### Solution

(i) The potential of the quinhydrone half-cell reaction measured against a standard hydrogen electrode is

$$E = 0.700 + \frac{RT}{F} \ln \left\{ \frac{a_{C_6H_4O_2}^{\frac{1}{2}} \cdot a_{H^+}}{a_{C_6H_4(OH)_2}^{\frac{1}{2}}} \right\}$$

For an equimolar mixture of benzoquinone and hydroquinone and assuming that their activity coefficients are similar

$$\frac{a_{C_6H_4O_2}}{a_{C_6H_4(OH)_2}} \approx 1$$

so

$$E = 0.700 + \frac{RT}{F} \ln a_{H^+}$$

But

$$pH = -\log_{10} a_{H^+}$$

$$= -\frac{\ln a_{H^+}}{\ln 10}$$

so that

$$E = 0.700 - \frac{2.303RT}{F} \cdot \text{pH}$$

Measurement of $E$ thus provides a direct measure of pH, scaled by the factor $(2.303RT/F) \approx 59$ mV.

(ii) If the standard hydrogen electrode is replaced by a SCE,

$$E = 0.460 - \frac{2.303RT}{F} \cdot \text{pH}$$

If the measured EMF is

$$E = +0.160 \text{ V}$$

then

$$0.300 = \frac{2.303RT}{F} \cdot \text{pH}$$

so that

$$\text{pH} = 5.1$$

(iii) The first acid dissociation constant of hydroquinone:

$$C_6H_4(OH)_2 \rightleftharpoons C_6H_4(OH)(O^-) + H^+$$

has a value of $K_{a1} = 1.2 \times 10^{-10}$ ($pK_{a1} = 9.9$). Thus for pH > 9 the response of the quinhydrone electrode is less than expected on the basis that the molecule is fully protonated and in the $C_6H_4(OH)_2$ form. The second dissociation constant is $pK_{a2} \simeq 17$ so that above pH 11 the following equilibrium will hold

$$\frac{1}{2}C_6H_4O_2 + \frac{1}{2}H^+ + e^- \rightleftharpoons \frac{1}{2}C_6H_4(OH)(O^-)$$

and

$$E = E^\ominus - \frac{2.303RT}{2F} \cdot \text{pH}$$

where

$$E^\ominus = 0.700 + \left\{ -\frac{2.303RT}{2F}pK_{a1} \right\}$$

## 1.26  Standard Potentials and pH

## Problem

Anthraquinone monosulphonate, A, is a reversible $2e^-$, $2H^+$ redox system. Use the Nernst equation to describe how the equilibrium potential will shift with pH in aqueous media below pH 7. This is below the $pK_a$s associated with the reduced form, such that the reaction always involves the transfer of two protons and two electrons.

## Solution

The reaction is

$$A + 2e^- + 2H^+ \rightleftharpoons AH_2$$

where $AH_2$ is the di-reduced di-protonated form of the anthraquinone species. Hence

$$E = E_f^{\ominus} - \frac{RT}{2F} \ln \frac{[AH_2]}{[A][H^+]^2}$$

$$pH = -\log a_{H^+} \simeq -\log[H^+]$$

$$E = E_f^{\ominus} + \frac{2RT}{2F} \ln[H^+] - \frac{RT}{2F} \ln \frac{[AH_2]}{[A]}$$

Then

$$E = E_{ref}^{\ominus} - \frac{RT}{2F} \ln \frac{[AH_2]}{[A]}$$

where

$$E_{ref}^{\ominus} = E_f^{\ominus} - \frac{RT}{F} \ln 10 \cdot pH$$

Therefore

$$\frac{\partial E}{\partial \, pH} = -\frac{RT}{F} \ln 10$$

$$= -0.059 \text{ V at } 298 \text{ K}$$

## 1.27  Standard Potentials and pH

## Problem

The electroreduction of anthraquinone monosulphonate, A, is a reversible $2H^+$, $2e^-$ process below pH 7.

(i) Derive a general expression to describe how the equilibrium potential for this process will vary across the full pH range (0–14). Note that this must recognise the two $pK_a$ values associated with the fully reduced form.

(ii) The equilibrium potential for this species has been measured as a function of pH (between pH 4 and 13), as given in the table below. Using your answer for (i) determine the $pK_a$ values associated with the reduced species (this is best achieved through the use of numerical fitting software).

| pH | 4 | 5.01 | 6.06 | 6.99 | 7.95 |
|---|---|---|---|---|---|
| $E_{eq}$ /V | −0.304 | −0.363 | −0.417 | −0.465 | −0.521 |

| pH | 9.01 | 9.99 | 11 | 11.91 | 12.73 |
|---|---|---|---|---|---|
| $E_{eq}$ /V | −0.554 | −0.586 | −0.605 | −0.614 | −0.611 |

## Solution

(i) The variation of the (reversible) equilibrium potential with pH is described by the Nernst equation. The following chemical and electrochemical steps must be considered:

$$A + 2e^- \rightleftharpoons A^{2-} \qquad\qquad E_1$$
$$AH^- \rightleftharpoons A^{2-} + H^+ \qquad K_1$$
$$AH_2 \rightleftharpoons AH^- + H^+ \qquad K_2$$

where $AH_2$ is the dihydroanthraquinone. Note that the acid–base equilibria have been written as deprotonations such that

$$K_1 = \frac{[A^{2-}][H^+]}{[AH^-]} \tag{1.16}$$

$$K_2 = \frac{[AH^-][H^+]}{[AH^2]} \tag{1.17}$$

In order to proceed, $[A^{2-}]$ needs to be expressed in terms of $[AH_2]_{tot}$, where

$$[AH_2]_{tot} = [A^{2-}] + [AH^-] + [AH_2] \tag{1.18}$$

Rearranging Eq. 1.18 and substituting in Eq. 1.17 gives

$$[A^{2-}] = [AH_2]_{tot} - [AH^-]\left(1 + \frac{[H^+]}{K_2}\right)$$

followed by substituting in Eq. 1.16 gives

$$[A^{2-}] = [AH_2]_{tot} - [A^{2-}]\left(\frac{K_2[H^+] + [H^+]^2}{K_1 K_2}\right)$$

Rearrangement leads to

$$[A^{2-}] = \frac{[AH_2]_{tot} K_1 K_2}{K_1 K_2 + K_2[H^+] + [H^+]^2} \tag{1.19}$$

The Nernst equation for the redox process is

$$E = E_1 - \frac{RT}{2F} \ln \frac{[A^{2-}]}{[A]}$$

Substitution of Eq. 1.19 into the Nernst equation leads to the following equation which describes the equilibrium potential of the redox couple across the full pH range:

$$E = E_1 - \frac{RT}{2F} \ln \frac{[AH_2]_{tot}}{[A]} K_1 K_2$$
$$+ \frac{RT}{2F} \ln \left(K_1 K_2 + K_2[H^+] + [H^+]^2\right) \tag{1.20}$$

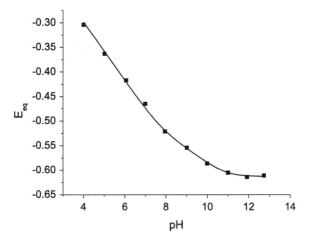

**Fig. 1.2** The variation of the equilibrium potential for the AQMS system as a function of pH. The line is the theoretical result given by Eq. 1.20. Adapted from C. Batchelor-McAuley *et al., J. Phys. Chem. B* **114** (2010) 4094, with permission from the American Chemical Society.

(ii)  In order to fit Eq. 1.20 to the experimental data given, it is necessary to assume that the activity coefficient for $H^+$ is unity so that

$$pH \approx -\log[H^+]$$

Figure 1.2 shows the best fit of Eq. 1.20 to the experimental data given, where the values of $K_1$ and $K_2$ are found to be $1.2 \times 10^{-11}$ and $2.0 \times 10^{-8}$, respectively.

# 2

---

# Electrode Kinetics

## 2.1 Faraday's Laws of Electrolysis

## Problem

The British physicist Michael Faraday was one of the principal early investigators of electrochemistry. His conclusions may be summarised in what are now known as Faraday's laws of electrolysis, which describe bulk electrochemical reactions:

1. The quantity of charge transferred at an electrode is directly proportional to the mass of material reacted at that electrode.
2. The constant of proportionality between charge quantity and reaction mass is itself proportional to the molar mass of the reactant so that, with the convention of reductive charge being negative:

$$m = \frac{-Q}{nF} M_r$$

where $m$ is the reaction mass in kg; $Q$ is the charge passed in C; $n$ is an integer constant; $F$ is described as the Faraday constant and has units $C\,mol^{-1}$; and $M_r$ is the molar mass of the reactant in $kg\,mol^{-1}$.

(i) Consider the process of bulk electrolysis in terms of the concepts of conservation of mass and conservation of charge. Explain the origin of Faraday's laws and define the constants $n$ and $F$.
(ii) How might the Faraday constant be measured experimentally?

## Solution

(i) Any electrochemical reaction may be written in terms an arbitrary reactant species A gaining $n$ electrons to form an arbitrary product B. $n$ is necessarily an integer as each molecule can only exchange discrete numbers of electrons:

$$A + ne^- \rightleftharpoons B$$

This is a balanced equation and defines $n$ as the number of electrons gained per molecule of A; if $n$ is negative, $-n$ is equally the number of electrons lost.

Therefore, if one mole of A reacts, then $n$ moles of electrons are consumed. The charge that is passed is then $-nF$, charge being negative for a reduction, when $F$ is defined as the charge on one mole of electrons. $F$ is the Faraday constant; it is equal to the Avogadro constant ($6.022 \times 10^{23}$ mol$^{-1}$) multiplied by the charge on a single electron ($1.602 \times 10^{-19}$ C), and has the value $\approx$ 96 485 C mol$^{-1}$. It is evident that $-Q/nF$ is the number of moles of A reacted, and so the mass reacted is related to this quantity by the molar mass of A.

(ii) The Faraday constant is calculated by measuring the charge passed when a known number of moles is reacted to completion. This is typically performed by electrodepositing silver onto an electrode under conditions of controlled current:

$$Ag^+_{(aq)} + e^- \rightleftharpoons Ag_{(s)}$$

Here, $n = 1$. If the current passed, $I$, is constant, then $Q = It$, with $t$ the time in seconds of deposition. Since the electrodeposited silver can be weighed, and the atomic weight of silver is known ($\approx$108 g mol$^{-1}$), $F$ is easily inferred from Faraday's laws of electrolysis.

## 2.2 Electrodeposition

### Problem

A British penny is manufactured by electroplating of copper onto mild steel. Figure 2.1 shows data for the deposition of Cu on a steel surface. The current, $I$, passed at the electrode is measured amperometrically and plotted against experiment time.

(i) How can the quantity of material deposited be calculated from such a plot?
(ii) A steel coin with diameter ($2r$) 2 cm and thickness ($h$) 1.5 mm is electroplated with copper using a solution of $Cu^{2+}$. A reductive charge of 29.4 C is passed. What thickness, $d$, of Cu has been deposited, given that the density of Cu is 8930 kg m$^{-3}$ and its molar mass is 63.5 g mol$^{-1}$?

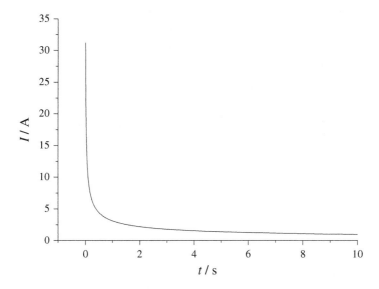

**Fig. 2.1** Electroplating of Cu onto a steel surface, measured amperometrically.

(iii) The underpotential deposition of silver from $Ag^+$ onto a thymine-modified gold $Au(111)$ surface was reported in the recent literature [A. Vollmer *et al.*, *J. Electroanal. Chem.* **605** (2007) 15]. The first deposition peak was integrated and it was determined that electrodeposition took place with $Q \approx 200\,\mu C\,cm^{-2}$. Given that the atomic radius of Ag is $\approx 144\,pm$, is this consistent with deposition of a single monolayer?

## Solution

(i) Charge and current are still related even where current is observed rather than controlled and so is not a constant. Current is defined:

$$I = \frac{\partial Q}{\partial t}$$

and so

$$Q(t) - Q(0) = \int_0^t I(t)\,dt$$

In an experimental situation, this integration can be performed numerically using a data analysis program. The quantity (in moles) of material deposited can then be deduced directly from the charge passed, provided $n$ is known.

(ii) From the charge passed, we can calculate the mass of Cu deposited:

$$m = \frac{-Q}{nF}M_r = \frac{29.4}{2 \times 96485} \times \frac{63.5}{1000} = 9.67 \times 10^{-6}\,kg$$

From the density we can determine the volume of Cu:

$$V = \frac{m}{\rho} = \frac{9.67 \times 10^{-6}}{8930} = 1.08 \times 10^{-9} \, \text{m}^3$$

so that dividing by the surface area, $A = 2\pi r^2 + 2\pi rh$, for a cylinder:

$$r_{Cu} = \frac{V}{A} = \frac{1.08 \times 10^{-9}}{2\pi \times 0.01 \times (0.01 + 0.0015)} = 1.5 \times 10^{-6} \, \text{m}$$

Hence the thickness of the electroplating is 1.5 $\mu$m.

(iii) Assuming that an Ag monolayer is close-packed and that the atoms are spherical and in contact, the centre of each atom is separated by 288 pm and the unit cell of one Ag atom is a rhombus with side 288 pm (Fig. 2.2).

Hence the area of one unit cell is $(288^2 \times (\sqrt{3}/2))$ pm$^2$, and so the area per molecule can be inferred:

$$\Gamma = \frac{2}{\sqrt{3}} \times \frac{1}{(288 \times 10^{-12})^2}$$

$$= 1.392 \times 10^{19} \, \text{moleculesm}^{-2}$$

Dividing by the Avogadro constant

$$\Gamma = 2.31 \times 10^{-5} \, \text{mol m}^{-2}$$

Since the reduction of Ag$^+$ to Ag is a one-electron process, the charge passed per unit area is found by multiplying by the Faraday constant, so that

$$Q = 2.31F \times 10^{-5} = 2.23 \, \text{C m}^{-2} = 223 \, \mu\text{C cm}^{-2}$$

Hence the observed $Q = 200 \, \mu\text{C cm}^{-2}$ is consistent with deposition of an ideal monolayer to within 11%, which is good agreement considering the crude geometric model and experimental error. It is therefore reasonable to assume that a single monolayer is deposited in this peak.

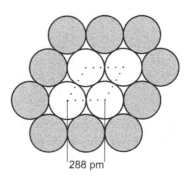

288 pm

**Fig. 2.2**  Close-packed monolayer of Ag (ideal geometry).

## 2.3 Tafel Analysis: One-Electron Processes

## Problem

For a fully electrochemically irreversible one-electron system, show how analysis of the voltammetry may yield information about the transition state for the process.

## Solution

For the process

$$A + e^- \rightleftharpoons B$$

the Butler–Volmer equation (as given below, Eq. 2.1) describes parametrically how the current associated with a redox process depends upon the electrode potential when the electron transfer is rate-limiting.

$$I = -FAk^0 \left( \exp\left[\frac{-\alpha F}{RT}(E - E_f^\ominus)\right] [A]_0 - \exp\left[\frac{(1-\alpha)F}{RT}(E - E_f^\ominus)\right] [B]_0 \right)$$

$$(2.1)$$

where $F$ is the Faraday constant, $A$ is the area of the electrode, $k^0$ is the *standard electrochemical rate constant* and $[i]_0$ designates the concentration of species $i$ at the electrode.

When $E \gg E_f^\ominus$ or $E \ll E_f^\ominus$, the Butler–Volmer equation reduces to the one-term Tafel equation as the first or second exponential term, respectively, may be approximated as zero. Consequently, a plot of $\ln|I|$ vs $E - E_f^\ominus$ for the region highlighted in Fig. 2.3 (often called the Tafel region) should yield a straight line of gradient $\alpha F/RT$, so allowing measurement of the transfer coefficient ($\alpha$). $\alpha$ (as shown in Eq. 2.1) is known as a transfer coefficient and is a measure of the position of the transition state between the oxidised and reduced species. Typically it has a value of around 0.5.

Note that the data nearer the voltammetric peak reflect both electrode kinetics and diffusional depletion, thus explaining the choice of data used to find $\alpha$.

## 2.4 Tafel Analysis: Electrochemically Reversible Processes

## Problem

For a fully reversible one-electron reduction, what gradient will be obtained for a plot of $\ln|I|$ vs $E$? Assume that the redox species is confined to a thin layer such that

$$[A]_0 + [B]_0 = [A]_{bulk} \qquad (2.2)$$

and that the diffusive flux may be treated as proportional to the concentration difference ($[A]_0 - [A]_{bulk}$).

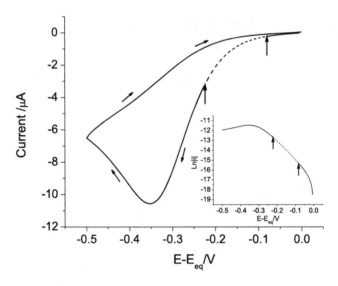

**Fig. 2.3** The cyclic voltammetric response for an irreversible one-electron reduction process, with the region required for Tafel analysis highlighted. The inset shows the Tafel plot for the forward scan highlighted is the required linear region. The voltage scan starts at 0.0 V and sweeps negatively to −0.5 V before returning to 0.0 V (small arrows indicate scan direction).

## Solution

If the reduction is reversible, the concentrations of species A and B are determined by the Nernst equation:

$$\frac{[A]_0}{[B]_0} = \exp(\theta) \tag{2.3}$$

where

$$\theta = \frac{F}{RT}\left[E - E^{\ominus}_{f,A/B}\right]$$

Substitution of Eq. 2.2 into Eq. 2.3, followed by rearrangement gives

$$[A]_0 = \frac{1}{1 + \exp(-\theta)}[A]_{bulk}$$

The current is given by

$$I = FAJ$$

$$\propto ([A]_0 - [A]_{bulk})$$

$$= [A]_{bulk}\frac{-1}{1 + \exp(\theta)}$$

where $A$ is the electrode area and $J$ is the diffusive flux. In the Tafel region (low overpotential, large $\theta$ for a reduction)

$$I \propto \frac{-1}{1 + \exp(\theta)}$$

$$\approx -\exp(-\theta)$$

so

$$\frac{\partial \ln|I|}{\partial \theta} \to -1$$

and hence in the Tafel region

$$\frac{\partial \ln|I|}{\partial E} = \frac{-F}{RT}$$

The gradient of the plot is then $-F/RT$ and the apparent $\alpha$ is unity, for a reversible process.

## 2.5 Tafel Analysis: Mass Transport Correction

### Problem

Using the Nernst diffusion layer model, derive an equation which describes how the flux ($J_A$) varies with overpotential for a one-electron reduction,

$$A + e^- \rightleftharpoons B$$

Assuming that the electron transfer is fully electrochemically reversible and that the diffusion layer thickness is constant at all potentials, what would the gradient be for a plot of $\ln[\frac{I_{\lim}}{I} - 1]$ vs $E$?

### Solution

The flux of A across the Nernst layer may be described by

$$J_A = D\frac{[A]_0 - [A]_{bulk}}{\delta} \tag{2.4}$$

where $[A]_0$ is the concentration of species A at the electrode surface, $[A]_{bulk}$ is the bulk concentration of A in solution, $\delta$ is the diffusion layer thickness and $D$ is the diffusion coefficient of species A.

We will assume the diffusion coefficients for species A and B to be equal. The law of conservation of mass then requires that

$$[A]_0 + [B]_0 = [A]_{bulk}$$

where $[B]_0$ is the concentration of B at the electrode surface. If, as may be experimentally expected, the concentration of B in the bulk solution is zero, then we may write

$$J_A = -J_B = -D\frac{[B]_0}{\delta} \tag{2.5}$$

From the Butler–Volmer equation we know,

$$J_A = -k_0(\exp[-\alpha\theta][A]_0 - \exp[(1-\alpha)\theta][B]_0) \tag{2.6}$$

Substitution of Eqs. 2.4 and 2.5 into Eq. 2.6 gives

$$J_A = \frac{-k_0\,\exp[-\alpha\theta][A]_{\text{bulk}}}{1 + \frac{\delta k_0}{D}\exp[-\alpha\theta] + \frac{\delta k_0}{D}\exp[(1-\alpha)\theta]} \tag{2.7}$$

Equation 2.7 may be simplified for a the case of a reversible electron transfer where $k_0\delta/D \gg 1$ so that,

$$J_A = -\frac{D[A]_{\text{bulk}}}{\delta(1 + \exp[\theta])} \tag{2.8}$$

It should also be noted that at high reductive overpotential, $\exp[(1-\alpha)\theta] \to 0$. This reduces Eq. 2.7 to that given below, and thus provides an expression for the limiting current

$$J_{A,\text{lim}} = -D\frac{[A]_{\text{bulk}}}{\delta} \tag{2.9}$$

Substituting Eq. 2.9 into Eq. 2.8 gives,

$$J_A = \frac{J_{A,\text{lim}}}{(1 + \exp[\theta])}$$

We know that $I = FAJ_A$ for a reduction, and hence a plot of $\ln[\frac{I_{\text{lim}}}{I} - 1]$ vs $E$ would yield a straight line of gradient $F/RT$.

## 2.6 Tafel Analysis: Two-Electron Processes

### Problem

The mechanism for a two-electron process is shown below:

$$A + e^- \rightleftharpoons B \qquad E_{f,A/B}$$
$$B + e^- \rightleftharpoons C \qquad E_{f,B/C}$$

In cases where the voltammetry of a two-electron process exhibits a single reductive wave ($E_{f,A/B} \leq E_{f,B/C}$), what gradient would be obtained from a Tafel plot ($\ln|i|$ vs $E$) in each of the following cases?

(i) The first electron transfer ($E_{f,A/B}$) is the rate-determining step.

(ii) The second electron transfer ($E_{f,B/C}$) is the rate-determining step, following a pre-equilibrium.

(iii) Both electron transfers are fully reversible.

Comment on your answers, highlighting why these are useful results for the mechanistic analysis of electrochemical systems.

## Solution

If the transfer coefficients ($\alpha_1, \beta_1$) and ($\alpha_2, \beta_2$) refer to the one-electron transfers from 'A to B' and 'B to C', respectively, the fluxes of all three species can be written as

$$J_A = -k_{A/B}^0 \exp\left[\frac{-\alpha_1 F}{RT}\left(E - E_{f,A/B}^\ominus\right)\right][A]_0$$

$$+ k_{A/B}^0 \exp\left[\frac{\beta_1 F}{RT}\left(E - E_{f,A/B}^\ominus\right)\right][B]_0$$

$$J_B = +k_{A/B}^0 \exp\left[\frac{-\alpha_1 F}{RT}\left(E - E_{f,A/B}^\ominus\right)\right][A]_0$$

$$- k_{A/B}^0 \exp\left[\frac{\beta_1 F}{RT}\left(E - E_{f,A/B}^\ominus\right)\right][B]_0$$

$$- k_{B/C}^0 \exp\left[\frac{-\alpha_2 F}{RT}\left(E - E_{f,B/C}^\ominus\right)\right][B]_0$$

$$+ k_{B/C}^0 \exp\left[\frac{\beta_2 F}{RT}\left(E - E_{f,B/C}^\ominus\right)\right][C]_0$$

$$J_C = +k_{B/C}^0 \exp\left[\frac{-\alpha_2 F}{RT}\left(E - E_{f,B/C}^\ominus\right)\right][B]_0$$

$$- k_{B/C}^0 \exp\left[\frac{\beta_2 F}{RT}\left(E - E_{f,B/C}^\ominus\right)\right][C]_0$$

So that the overall current for the process is

$$I = -FA(J_B + 2J_C)$$

From conservation of mass

$$J_A + J_B + J_C = 0$$

hence,

$$I = -FA(J_C + (J_B + J_C))$$
$$I = -FA(J_C - J_A)$$

(i) When the first step is rate-determining, all B produced will be immediately consumed; hence $[B]_0 = 0$. Further, the rate of production of C will be opposite and equal to the rate of loss of A ($J_A = -J_C$):

$$I = -FA\,(-2J_A)$$

$$= -2FA\,k^0_{A/B}\exp\left[\frac{-\alpha_1 F}{RT}\left(E - E^{\ominus}_{f,A/B}\right)\right][A]_0$$

Tafel analysis yields the following result,

$$\ln|I| = -\frac{\alpha_1 FE}{RT} + \text{constant}$$

(ii) For the case where the second step is rate-determining and irreversible, by assuming a pre-equilibrium, $A \rightleftharpoons B$, we can say that the flux of A is negligible ($j_A \approx 0$) so that:

$$I = -FA\,(J_C)$$

$$= -FA\,k^0_{B/C}\exp\left[\frac{-\alpha_2 F}{RT}\left(E - E^{\ominus}_{f,B/C}\right)\right][B]_0$$

From the Nernst equation we know

$$\frac{[B]_0}{[A]_0} = \exp\left[\frac{-F}{RT}\left(E - E^{\ominus}_{f,A/B}\right)\right]$$

hence,

$$I = -FA\,k^0_{B/C}\exp\left[\frac{-\alpha_2 F}{RT}\left(E - E^{\ominus}_{f,B/C}\right)\right]$$

$$\times \exp\left[\frac{-F}{RT}\left(E - E^{\ominus}_{f,A/B}\right)\right][A]_0$$

So that

$$\ln|I| = -\frac{(1+\alpha_2)FE}{RT} + \text{constant}$$

(iii) When both electron transfers are fully reversible we can extend the answer given in Problem 2.4 to account for more than one electron. From the Nernst equation we have

$$[C] = [A]\exp\left[\frac{-2F}{RT}\left(E - E^{\ominus}_f\right)\right]$$

Analogously to Problem 2.4

$$I \propto \exp\left(\frac{-2F}{RT}\left(E - E_f^{\ominus}\right)\right)$$

As such a plot of $\ln |I|$ vs $E$ will yield a straight line of gradient $-2F/RT$, in the Tafel region.

The above results demonstrate how Tafel analysis provides a readily available method to assess which electron transfer is rate-limiting. More generally, the gradient of the plot is proportional to $n' + \alpha_{rds}$ where $n'$ is the number of electrons transferred prior to the rate-limiting step and $\alpha_{rds}$ is the transfer coefficient associated with the rate-limiting electron transfer. A simple but descriptive example of such a system is found with the hydrogen evolution reaction, where the rate-determining step changes with the electrode material used (see Problem 2.8). For more information on the theory behind Tafel analysis see the work of S. Fletcher [*J. Solid State Electrochem.* **13** (2009) 537].

## 2.7 The Butler–Volmer Equation and the Nernst Equation

## Problem

Show how the Butler–Volmer equation (as given below) reduces to the Nernst equation for a reversible one-electron process:

$$I = FAk^0\left(\exp\left[\frac{-\alpha F}{RT}\left(E - E_f^{\ominus}\right)\right][A]_0 - \exp\left[\frac{\beta F}{RT}\left(E - E_f^{\ominus}\right)\right][B]_0\right)$$

where $I$ is the current at a uniformly accessible macroelectrode, $F$ is the Faraday constant, $A$ is the area of the electrode, $k^0$ is the standard electrochemical rate constant and $[i]_0$ is the concentration of species $i$ at the electrode surface.

## Solution

In the case of a reversible electrochemical process the standard electrochemical rate constant ($k^0$) is large, such that $FAk^0 \gg I$, and therefore $I/FAk^0 \simeq 0$. Consequently,

$$\exp\left[\frac{-\alpha F}{RT}(E - E_f^{\ominus})\right][A]_0 = \exp\left[\frac{\beta F}{RT}(E - E_f^{\ominus})\right][B]_0$$

Note that for a one-electron transfer, $\alpha + \beta = 1$. Rearrangement then leads to the Nernst equation as given below:

$$\frac{[B]_0}{[A]_0} = \exp\left[\frac{-F}{RT}(E - E_f^{\ominus})\right]$$

## 2.8 The Hydrogen Evolution Reaction

### Problem

A proposed mechanism for the hydrogen evolution reaction $(2H^+ + 2e^- \rightarrow H_2)$ is given below,

$$H^+_{(aq)} + e^- \rightleftharpoons H^{\cdot}_{(ads)} \tag{2.10}$$

$$H^{\cdot}_{(ads)} + H^+_{(aq)} + e^- \rightleftharpoons H_{2(ads)} \tag{2.11}$$

or

$$2H^{\cdot}_{(ads)} \rightleftharpoons H_{2(ads)} \tag{2.12}$$

$$H_{2(ads)} \rightleftharpoons H_{2(g)} \tag{2.13}$$

Using this mechanism, explain why the current density for the reduction of $H^+$ varies widely with the nature of the metal electrode. For example, the rate on a Pt electrode is nine orders of magnitude higher than on a Hg electrode, and five orders of magnitude higher than on a Ta electrode. Explain these observations.

### Solution

Problem 2.6 demonstrated that for the Tafel analysis of a multi-electron system we expect to measure a gradient which is proportional to $n' + \alpha_{RDS}$, where $n'$ is the number of electrons transferred prior to the rate-determining step and $\alpha_{RDS}$ is the transfer coefficient for the rate-determining step, note that if all electron transfers are reversible and highly driven then the gradient is proportional to $n$ (the number of electrons).

 (i) Case A (Hg): the first electron transfer is the rate-determining step (Eq. 2.10). Such systems yield a Tafel plot with a gradient of $\simeq 0.5$. Further reaction of the $H^{\cdot}_{(ads)}$ species may proceed via either Eqs. 2.11 or 2.12.
 (ii) Case B (Pt, Ta): the second electron transfer is the rate-determining step (Eq. 2.11). Such systems yield a Tafel plot of gradient $\simeq 1.5$.
 (iii) Case C: here, $\alpha$ is found to be $\simeq 2$. This case is specific to palladium and is found where the rate-determining step is Eq. 2.12.

The rates of electron transfer associated with each metal and the cause of the change in mechanism may be understood through studying the estimated enthalpy of adsorption of $H^{\cdot}$ on the metal surfaces.

For metals upon which the adsorption of $H^{\cdot}$ is weak, Eq. 2.10 is the rate-determining step. As the strength of the binding increases the rate of the first electron transfer increases and hence the current density increases. As the strength

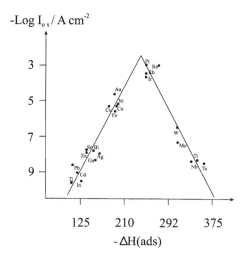

**Fig. 2.4** Current density for electrolytic hydrogen evolution versus strength of intermediate metal-hydrogen bond formed during the electrochemical reaction. Reprinted from S. Trasatti, *J. Electroanal. Chem.* **39** (1972) 163, with permission from Elsevier.

of the bonding increases further, Eq. 2.11 becomes the rate-determining step and further increases lead to a decrease in the overall rate of electron transfer. This information is well summarised in a 'volcano plot' as shown in Fig. 2.4.

## 2.9 Requirement for Supporting Electrolyte

### Problem

(i) Discuss the necessity of supporting electrolyte for conventional voltammetry, with reference to: a) the double layer; b) electric fields in bulk solution; c) the non-ideality of an electrolytic solution.

(ii) Under what conditions might the addition of excess supporting electrolyte be inappropriate or impossible?

### Solution

(i) Supporting electrolyte is conventionally added to a solution when performing voltammetry in order to elevate the conductivity of the solution and to suppress electric fields. This is particularly important at the interface with the electrode where electron transfer actually takes place.

　　The charge placed on the electrode to drive a potential difference with respect to the solution is compensated by a double layer in which there is an excess of ions of opposite charge, as is discussed in more detail in Chapter 10.

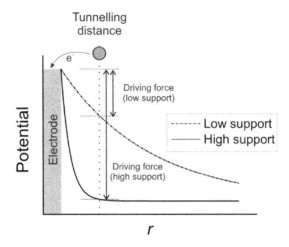

**Fig. 2.5** Different potential distributions at varying supporting electrolyte concentration, showing the limited driving force at weak support (low electrolyte concentrations).

This double layer is narrower for a more conductive solution, allowing electron tunnelling to take place from outside the double layer, and thus ensuring that the full potential difference applied between the working and reference electrodes is 'felt' by a molecule approaching the working electrode. The driven Faradaic current then reflects the applied potential.

Figure 2.5 demonstrates the difference in driven potential between a well-supported and weakly supported solution. The consequent difference in the perceived rate of reaction is given by the Frumkin correction as calculated from the difference in driving forces (indicated in the figure), as well as altered concentrations in the double layer. The derivation is discussed in Problem 2.10.

The screening of bulk electric fields by the addition of supporting electrolyte is useful since it prevents migration due to electrostatic attractions from influencing the voltammetric current and allows the use of a simpler 'diffusion-only' theory based on Fick's laws (see Chapter 3). What is more, passing current through a resistive solution generates a potential difference known as 'ohmic drop', since its value can be understood from Ohm's law:

$$\Delta\phi_{OD} = IR_s$$

where $\Delta\phi_{OD}$ is ohmic drop, $I$ is the driven current and $R_s$ is the solution resistance. By lowering the solution resistance with supporting electrolyte, $R_s$ is minimised and therefore the ohmic drop remains small. Where ohmic drop is large, further overpotential is required to drive a given current, thereby distorting voltammetric features (see Chapter 10).

Lastly, the uniformity of ionic strength provided to a solution by the presence of ample supporting electrolyte limits effects due to the non-ideality of the solution. According to the Debye–Hückel theory, the presence of electrostatic interactions between ions causes solution non-ideality because these forces are on average stabilising. Therefore, activity, the quantity appearing in the Nernst equation, differs from concentration by a factor known as the activity coefficient, $\gamma_i$, in a manner which for a dilute ($<0.01$ M) solution is given by a simplified formula:

$$\log \gamma_i \approx -A z_i^2 \sqrt{I}$$

where $A$ is a characteristic coefficient, $z_i$ is the ionic charge number and $I$ is here the ionic strength.

The addition of supporting electrolyte generally elevates concentrations to regimes where the Debye–Hückel formula does not apply, but it also ensures that non-ideality is uniform across the solution and therefore gradients of non-ideality do not contribute to mass transport. Note that because a well-supported solution is non-ideal, the exact value of the formal potential, $E_f^{\ominus}$, of a redox couple is sensitive to the concentration of supporting electrolyte.

(ii) The addition of supporting electrolyte is problematic for the voltamme-try of biological molecules, since such molecules are typically sensitive in their conformation and reactivity to salt concentration. Additionally, in cer-tain less polar solvents it may be impossible to dissolve adequate supporting electrolyte to avoid Frumkin corrections (see Problem 2.10) and fully sup-press ohmic drop. Lastly, as is discussed in Chapter 10, the addition of sup-porting electrolyte sometimes masks interesting voltammetric features that may provide mechanistic insights beyond those available from diffusion-only voltammetry.

## 2.10 Frumkin Corrections

### Problem

Suppose that due to insufficient supporting electrolyte being present, an uncom-pensated potential difference $\Delta\phi_{PET}$ exists between the plane of electron transfer (PET) and the outer edge of the double layer. Suppose additionally that within the double layer, the concentration of the electroactive species A is equilibrated, and hence obeys the Boltzmann equation:

$$c_A = c_{A,0} \exp\left(-\frac{z_A F}{RT}(\phi - \phi_0)\right)$$

with reference to the concentration $c_{A,0}$ and potential $\phi_0$ at the outer edge of the double layer.

(i) If A undergoes an irreversible one-electron reduction according to a Tafel law, how will this equation be modified by $\Delta\phi_{PET}$, considering both the change in perceived overpotential and the change in the concentration of A at the PET?

(ii) Hence determine an expression for the observed heterogeneous rate constant $k_{app}^0$ as a function of the actual $k^0$ and of $\Delta\phi_{PET}$.

(iii) $k^0$ has been *measured* to be 0.1 cm s$^{-1}$ but $\Delta\phi_{PET}$ is estimated to be +50 mV. Given that $\alpha = 0.5$ and $z_A = 1$, apply the Frumkin correction formula you have determined to estimate the true value of $k^0$.

## Solution

(i) A generic Tafel law is:

$$J_A = k^0 \exp\left(-\alpha\theta\right) c_{A,PET}$$

Now the overpotential is altered as

$$\theta = \frac{F}{RT}(E - E_f^\ominus - \Delta\phi_{PET})$$

as indicated by the difference in driving forces in Fig. 2.5. The concentration at the PET can be expressed using the Boltzmann equation as

$$c_{A,PET} = c_{A,0} \exp\left(-\frac{z_A F}{RT}\Delta\phi_{PET}\right)$$

Therefore

$$J_A = k^0 \exp\left(-\frac{\alpha F}{RT}(E - E_f^\ominus - \Delta\phi_{PET})\right) \times \exp\left(-\frac{z_A F}{RT}\Delta\phi_{PET}\right) c_{A,0}$$

(ii) The apparent $k^0$ is that given by the Tafel law on the assumption that $\Delta\phi_{PET} = 0$:

$$J_A = k_{app}^0 \exp\left(-\frac{\alpha F}{RT}(E - E_f^\ominus)\right) c_{A,0}$$

By comparing the two expressions, it is clear that

$$k_{app}^0 = \exp\left(\frac{\alpha F}{RT}\Delta\phi_{PET}\right)\exp\left(-\frac{z_A F}{RT}(\phi - \phi_0)\right) k_0$$

which simplifies to:

$$k_{app}^0 = \exp\left((\alpha - z_A)\frac{F}{RT}\Delta\phi_{PET}\right) \cdot k^0$$

(iii) Substituting in the values given, we find:

$$k^0 \approx k^0_{\text{app}} \exp\left((z_A - \alpha)\frac{F}{RT}\Delta\phi_{\text{PET}}\right)$$

$$= 0.1 \times \exp\left((1 - 0.5) \times 38.94 \times 0.05\right)$$

$$= 0.1 \times \exp(0.974)$$

$$= 0.265 \text{ cm s}^{-1}$$

## 2.11 Marcus Theory and Standard Electrochemical Rate Constants

### Problem

Describe the basic principles involved in Marcus theory and hence rationalise qualitatively the heterogeneous *standard electrochemical rate constants* ($k^0$) for the following one-electron reductions of metal ions in aqueous solution:

| Species | $k^0$/m s$^{-1}$ (25°C) |
|---|---|
| $Ru(NH_3)_6^{3+}$ | $10^{-2}$ |
| $V^{3+}$ | $4 \times 10^{-5}$ |
| $Co(NH_3)_6^{3+}$ | $5 \times 10^{-10}$ |

### Solution

From Marcus theory we know that electron transfer proceeds through the thermal activation of a reactant to a 'transition state'. This 'transition state' ($\ddagger$) is depicted in Fig. 2.6 and shows it to be the point at which the two potential energy curves for the reactant (R) and product (P) cross over.

Once the molecule is at the cross-over point it is possible for electron tunnelling to take place and the product is formed in a highly excited vibrational level. The amount of energy required to reach this transition state ($\Delta G(\ddagger)$) controls the rate of electron transfer as given by Eq. 2.14,

$$k^0 = \kappa_{\text{el}} K_p \nu_N \exp\left[\frac{-\Delta G(\ddagger)}{RT}\right] \tag{2.14}$$

where $K_p$ is the equilibrium constant, $\nu_N$ is a nuclear frequency factor and $\kappa_{\text{el}}$ is the electronic transmission coefficient ($\kappa_{\text{el}} = 1$ for an adiabatic process). $\Delta G(\ddagger)$ has two major contributions,

$$\Delta G(\ddagger) = \Delta G_{\text{inner}}(\ddagger) + \Delta G_{\text{outer}}(\ddagger)$$

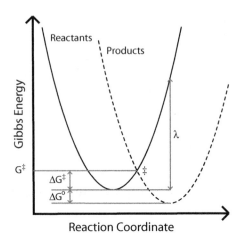

**Fig. 2.6** Schematic for the potential energy curves for an electrochemical reaction.

where $\Delta G_{\text{inner}}(\ddagger)$ is the activation energy arising from the distortion of the inner coordination shell and $\Delta G_{\text{outer}}(\ddagger)$ arises from the energy required to rearrange the dipoles associated with other solvent molecules. With this knowledge we can qualitatively explain the observed variation in rate constants for the metal complexes:

(i) $Ru(NH_3)_6^{3+}$

The ion exhibits a high electrochemical rate constant, hence suggesting that there is little rearrangement of either the outer (water) or inner ($NH_3$) co-ordination spheres on reduction.

(ii) $V^{3+}$

Vanadium is a first row transition metal. Consequently, its d-orbitals are more contracted, leading to the ion being more ionic in character than $Ru(NH_3)_6^{3+}$. This ionic character means there is a relatively large change in the orientation of the solvent molecules (outer-sphere) upon reduction, and, consequently, the electrochemical rate constant is smaller.

(iii) $Co(NH_3)_6^{3+}$

The $Co(NH_3)_6^{3+}$ ion is a low spin $d^6$ complex. Reduction leads to a high spin $d^7$ complex — such a transition is spin-forbidden and hence the rate of electron transfer is far lower.

## 2.12 Marcus Theory and Butler–Volmer Kinetics

### Problem

The Gibbs energy change between the reactant and the transition state, $\Delta G(\ddagger)$, (as discussed in Problem 2.11) may be described in terms of the reorganisation energy $\lambda$:

$$\Delta G(\ddagger) = \frac{\lambda}{4}\left(1 + \frac{\Delta G}{\lambda}\right)^2 \tag{2.15}$$

where the quantities $\Delta G(\ddagger)$, $\Delta G$ and $\lambda$ are defined in Fig. 2.6. In terms of $\alpha$ (the transfer coefficient), describe the relationship between Marcus theory and Butler–Volmer kinetics for a one-electron reduction.

## Solution

For a one-electron reduction, $\Delta G = F(E - E_f^\ominus)$, and the rate of electron transfer $(k)$ is given in terms of $\Delta G(\ddagger)$ as shown below:

$$k = a \exp\left(\frac{-\Delta G(\ddagger)}{RT}\right)$$

where $a$ is a constant.

We know from Butler–Volmer kinetics that

$$\alpha = -\frac{RT}{F}\frac{\partial \ln |I|}{\partial E}$$

$$= -\frac{RT}{F}\frac{\partial \ln |k|}{\partial E}$$

$$= \frac{1}{F}\frac{\partial \Delta G(\ddagger)}{\partial E}$$

where in the Tafel regime, $\Delta G(\ddagger) \propto \alpha F E$.

Hence from substitution of the expression for $\Delta G$ into Eq. 2.15

$$\Delta G(\ddagger) = \frac{\lambda}{4}\left(1 + \frac{F(E - E_f^\ominus)}{\lambda}\right)^2$$

such that on differentiation

$$\alpha = \frac{1}{F}\frac{\partial \Delta G(\ddagger)}{\partial E}$$

$$= \frac{1}{2}\left(1 + \frac{F(E - E_f^\ominus)}{\lambda}\right)$$

$$= \frac{1}{2}\left(1 + \frac{\Delta G}{\lambda}\right)$$

This result is important: it shows us that $\alpha$ is predicted by Marcus theory to be $\simeq 0.5$ for a one-electron process if $\lambda \gg \Delta G$, i.e. if the reaction is irreversible. Of further importance is the fact that $\Delta G$ as discussed above is not a standard Gibbs energy since it is dependent upon the electrode potential. Consequently, '$\alpha$' is not constant and this may lead to curved Tafel plots if $\alpha$ is measured over a wide potential range.

## 2.13 Marcus Theory and the Role of Solvent

### Problem

A high-speed channel electrode was used to measure the rates of electron transfer ($k^0$) for 9,10-diphenylanthracene (DPA) in a variety of solvents. The oxidation of DPA is an outer-sphere one-electron process. It was found that the measured values of $k$ were influenced by the reorientation dynamics of the solvent.

Below is a table which shows the measured rate constants for the oxidation and the longitudinal dielectric relaxation times ($t_L$) for the solvents [A.D. Clegg *et al.*, *J. Am. Chem. Soc.* **126** (2004) 6185].

(i) Explain what is meant by 'outer-sphere heterogeneous electron transfer'.

(ii) How will $v_N$ (the nuclear frequency factor) vary with $t_L$ for an outer-sphere electron transfer, assuming that the rearrangement of the inner sphere is negligible and the electronic coupling is small between the reactant and the electrode?

(iii) Suggest a suitable plot which will show that the data below are in agreement with your proposed answer for part (ii).

| Solvent | $k^0$/cm s$^{-1}$ (25°C) | $t_L$ /ps |
|---------|--------------------------|-----------|
| MeCN    | $0.94 \pm 0.16$          | 0.20      |
| EtCN    | $0.61 \pm 0.12$          | 0.31      |
| PrCN    | $0.32 \pm 0.07$          | 0.52      |
| BuCN    | $0.23 \pm 0.03$          | 0.74      |

### Solution

(i) In an outer-sphere electron transfer reaction, the integrity of the ion and its solvation shell is maintained and so there is no direct interaction between the ion and the electrode surface. This is in contrast with inner-sphere electron transfer, where the ion penetrates the outer Helmholtz plane and is in direct contact with the electrode surface.

(ii) The rate of electron transfer for an outer-sphere electron transfer is governed by

$$k^0 = \kappa_{el} K_p v_N \exp\left[\frac{-\Delta G^{\ddagger}}{RT}\right] \tag{2.16}$$

where $k^0$ is the standard rate constant, $\kappa_{el}$ is the electronic transmission coefficient, $v_N$ is the nuclear frequency factor and $\Delta G(\ddagger)$ is the energy required to reach the transition state. The nuclear frequency factor represents the rate at which reacting species close to the transition state are converted into products. This has contributions from bond vibrations and solvent motions. However, when the energy barrier for rearranging the inner sphere is small, the dynamics of the solvent will dominate:

$$\nu_N = t_L^{-1}\left(\frac{\Delta G^{\ddagger}}{4\pi RT}\right)^{\frac{1}{2}}$$

(iii) From the discussion above it can be seen that $k^0$ is proportional to $t_L^{-1}$; hence, a plot of $k^0$ versus $t_L^{-1}$ yields a straight line.

## 2.14 Marcus Theory and the Inverted Region

### Problem

(i) With reference to Marcus theory, what is the 'inverted region'?
(ii) It is not possible to observe the inverted region with heterogeneous electron transfer at a metallic electrode. Why?
(iii) What type of electrode materials may exhibit an inverted region?

### Solution

(i) For homogeneous electron transfers the variation in $\Delta G(\ddagger)$ is described by

$$\Delta G(\ddagger) = \frac{\lambda}{4}\left(1 + \frac{\Delta G}{\lambda}\right)^2$$

As shown in Eq. 2.16, the rate of electron transfer is dependent upon this value of $\Delta G(\ddagger)$. As the driving force for the reaction increases (an increase in $\Delta G$) the rate of electron transfer increases up to a maximum where $\Delta G = -\lambda$. At even more negative values of $\Delta G$ the activation energy increases and hence the rate of electron transfer decreases. This decrease in the rate of electron transfer with increasing driving force (i.e. increasingly negative $\Delta G$) is known as the 'inverted region'.

(ii) It is not possible to observe the inverted region for heterogeneous electron transfer at a metallic electrode. This is due to the fact that the above discussions on Marcus theory (Problems 2.11 and 2.12) are based on the assumption that the electron is being transferred to and from a narrow range of states (from the electrode's Fermi level to the molecule). Metallic electrodes have a wide distribution of states from which an electron may be transferred. Consequently, it is possible for an electron to be transferred from an electronic state below the Fermi level, so the inverted region is not observed.

An alternative approach, the Marcus–Hush formalisation, is based upon assessing the overlap of the electronic states in the electrode and the molecule present in solution, and accurately predicts that for high overpotentials a limitation in the rate of electron transfer ($k$) will be observed at a metallic electrode. For more information, see S.W. Feldberg [*Anal. Chem.* **82** (2010) 5176].

(iii) The electronic structure of semiconductors is markedly different from that of metals. Semiconductors exhibit band gaps such that there is a zero density of states at the Fermi level. The application of a potential to a semiconducting electrode may result in the population of the conducting band with electrons. These electrons are restricted to a narrow range of states, leading to a donor-acceptor type situation, whereby it is possible to observe the inverted region. Recent work by N.S. Lewis *et al.* [*Chem. Phys.* **326** (2006) 15] has experimentally demonstrated that it is possible to observe the inverted region on ZnO electrodes.

# 3

## Diffusion

## 3.1 Fick's Laws of Diffusion

### Problem

Fick's second law for one linear dimension $(x)$ is

$$\frac{\partial c}{\partial t} = D\frac{\partial^2 c}{\partial x^2}$$

where $x$ is distance, $t$ is time and $D$ is the diffusion coefficient of the diffusing species. If electrolysis leads to a concentration $c_0$ of a species at an electrode surface (located at $x = 0$) and the bulk concentration of the species becomes zero at a distance $x = \delta$ from the electrode surface, show that under steady-state conditions the concentration decreases linearly away from the electrode, and determine an equation for the steady-state distribution of $c$.

### Solution

At steady state

$$\frac{\partial c}{\partial t} = 0$$

so

$$D\frac{\partial^2 c}{\partial x^2} = 0$$

or

$$\frac{\partial^2 c}{\partial x^2} = 0$$

Integrating

$$\frac{\partial c}{\partial x} = A$$

where $A$ is a constant independent of $x$; hence the concentration profile is linear. Integrating again

$$c = Ax + B$$

where $B$ is another constant. When $x = 0$, $c = c_0$ so that

$$B = c_0$$

Also when $x = \delta$, $c = 0$ so that

$$A = -\frac{c_0}{\delta}$$

Hence

$$c = c_0\left(1 - \frac{x}{\delta}\right)$$

and the concentration of $c$ changes linearly from $c_0$ to zero between $x = 0$ and $x = \delta$. This model is the basis of the Nernst diffusion layer, where the concentration of the electroactive species is presumed to be maintained at or near its bulk solution value for $x \geq \delta$ due to natural convection.

## 3.2  Fick's Laws of Diffusion

### Problem

The concentration of a species undergoing steady-state electrolysis at an electrode is often approximated as changing linearly between the concentration at the electrode surface ($x = 0$, $c = c_0$) and the bulk concentration value at the edge of a diffusion layer ($x = \delta$, $c = c^*$) so that

$$c = c_0 + (c^* - c_0)\frac{x}{\delta} \quad 0 < x < \delta$$

Assuming steady-state conditions:

(i)  Show that the above expression satisfies Fick's second law, and
(ii)  Determine an expression for the flux to the electrode.

### Solution

(i)  Fick's second law shows that at steady state

$$\frac{\partial c}{\partial t} = 0 = D\frac{\partial^2 c}{\partial x^2}$$

From the expression given, differentiating

$$\frac{\partial c}{\partial x} = \frac{(c^* - c_0)}{\delta}$$

and differentiating once more

$$\frac{\partial^2 c}{\partial x^2} = 0$$

as required to satisfy Fick's second law.

(ii) Fick's first law states

$$J = -D \frac{\partial c}{\partial x}$$

so that

$$J = -D \frac{(c^* - c_0)}{\delta}$$

where the negative sign implies that the flux is in the direction of decreasing $x$, and hence towards the electrode. Therefore both increased bulk concentration and decreased electrode surface concentration will increase the flux, as would be expected.

## 3.3 Diffusion Distances

### Problem

The statistical (Einstein) view of diffusion is a random walk process, which suggests that the root-mean-square distance diffused by a species in time $t$ is

$$d \approx \sqrt{2Dt}$$

where a small molecule in aqueous solution typically has a diffusion coefficient $D \approx 10^{-5}$ cm$^2$ s$^{-1}$. Approximately how far would such a molecule diffuse in (i) one second and (ii) one day? Comment on any implications for electrochemical experiments.

### Solution

(i) For $D \approx 10^{-5}$ cm$^2$ s$^{-1}$ and $t = 1$ s,

$$\sqrt{2Dt} \approx 5 \times 10^{-3} \text{ cm}$$

$$\approx 50 \, \mu\text{m}$$

(ii) For $D \approx 10^{-5}$ cm$^2$ s$^{-1}$ and $t = 24 \times 60 \times 60$ s,

$$\sqrt{2Dt} = 1.3 \text{ cm}$$

The implication is that diffusion is a slow process in solution. Thus if bulk electrolysis is attempted, stirring or other forms of convection will be needed to ensure rapid and efficient conversion of the cell contents. Another implication is that with voltammetric experiments lasting a few seconds, the electrolysis is confined to a spatial layer of solution adjacent to the electrode, of the order of tens of microns in size.

## 3.4 The Cottrell Equation

### Problem

A potential step experiment was carried out in a solution containing 0.05 M ferrocyanide ($[Fe(CN)_6]^{4-}$) dissolved in a solution containing a large excess of inert electrolyte. Care was taken to ensure that there was no stirring of the solution during the experiment. The potential was stepped from a value where there was no reaction to a potential at which the $[Fe(CN)_6]^{4-}$ was oxidised to $[Fe(CN)_6]^{3-}$ at a mass transport controlled rate, and the following currents were recorded:

| $t/s$ | 0.1 | 0.2 | 0.4 | 0.8 | 1.2 |
|-------|-----|-----|-----|-----|-----|
| $I/mA$ | 6.9 | 4.9 | 3.4 | 2.4 | 2.0 |

(i) Explain why it is necessary to have a large excess of inert electrolyte present for the experiment.

(ii) Why is it important to make sure that there is no stirring of the solution during the experiment?

(iii) Make a sketch of the concentration of $[Fe(CN)_6]^{4-}$ as a function of distance away from the electrode immediately before the potential step and at two different times after the potential step.

(iv) Given that the area of the electrode was 0.3 cm², calculate the diffusion coefficient for $[Fe(CN)_6]^{4-}$ in the solution.

(v) Why is the time scale of the experiment limited to around a second?

### Solution

(i) The presence of the supporting electrolyte suppresses electric fields and ensures that the mass transport in the solution is exclusively via diffusion.

(ii) The absence of stirring is ensured to allow the current-time response to be described by a purely diffusional model, specifically the Cottrell equation:

$$I = nFAc^* \sqrt{\frac{D}{\pi t}}$$

where $D$ is the diffusion coefficient and $c^*$ the bulk concentration.

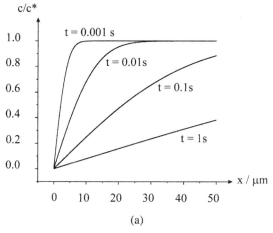

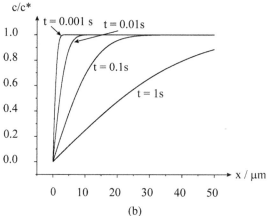

**Fig. 3.1** Concentration profiles at varying times after a potential step into the diffusion-controlled region (a) $D = 5 \times 10^{-5}$ cm$^2$ s$^{-1}$ and (b) $D = 5 \times 10^{-6}$ cm$^2$ s$^{-1}$. Reproduced from R.G. Compton *et al.*, *Understanding Voltammetry*, 2nd ed., with permission from Imperial College Press.

(iii) The concentration profiles are shown in Fig. 3.1 for $D = 5 \times 10^{-5}$ cm$^2$ s$^{-1}$ and $5 \times 10^{-6}$ cm$^2$ s$^{-1}$, and times of 0.001, 0.01, 0.1 and 1 s.

(iv) The Cottrell equation suggests that the data be analysed by plotting a graph of $I$ vs $1/\sqrt{t}$ which should, and does, give a straight line through the origin of

$$\text{gradient} = nFAc^* \sqrt{\frac{D}{\pi}} = 2.2 \times 10^{-3} \text{ A s}^{\frac{1}{2}}$$

where for the one-electron oxidation of ferricyanide, $n = 1$. Accordingly

$$D = 7.2 \times 10^{-6} \text{ cm}^2 \text{ s}^{-1}$$

(v)  After tens of seconds, natural convection will develop so that transport will no longer be purely diffusional and the Cottrell equation will become inaccurate. For this reason, only one second of experimental data is considered.

## 3.5  Derivation of the Cottrell Equation

### Problem

The Cottrell equation was developed in 1903 by solving a simple model system for chronoamperometry at a large electrode and at high overpotential [F.G. Cottrell, *Z. Physik. Chem.* **42** (1903) 385].

The equation can be solved by the method of Boltzmann transformation, as in *Understanding Voltammetry*, or alternatively by a mathematical method known as Laplace transformation, in which an integral transform is used to convert a partial differential equation into an ordinary differential equation. The transformation is:

$$\mathcal{L}\{f(t)\} = \bar{f}(s) = \int_0^\infty f(t) \cdot \exp(-st)\, dt$$

(i)  Show that:

   (a)  $\mathcal{L}\{k\} = k/s$, where $k$ is constant.
   (b)  $\mathcal{L}\{\partial f/\partial t\} = s\bar{f} - f(0)$.
   (b)  $\mathcal{L}\{\partial^2 f/\partial x^2\} = \partial^2 \bar{f}/\partial x^2$.

(ii)  Hence solve the Cottrell problem, which is to solve the diffusion equation under the conditions:

$$c(x,0) = c^*$$

$$c(\infty, t > 0) = c^*$$

$$c(0, t > 0) = 0$$

and determine the corresponding diffusion-limited current. Note that

$$\mathcal{L}\{\mathrm{erf}(a/2\sqrt{t})\} = \frac{1}{s}(1 - \exp(-a\sqrt{t})) \tag{3.1}$$

(iii)  How would the solution to part (ii) be changed if we require that the Nernst equation relates concentrations of reactant and product (reversible electron transfer), rather than assuming total reactant depletion? You may treat all diffusion coefficients as equal.

(iv)  Hence calculate the overpotential required for the Cottrell equation to be accurate to <1% and <0.1%.

## Solution

(i) (a)

$$\mathcal{L}\{k\} = \int_0^\infty k \cdot \exp(-st)\, dt$$

$$= k \int_0^\infty \exp(-st)\, dt$$

$$= k[-(1/s)\exp(-st)]_0^\infty$$

$$= k/s$$

(b)

$$\mathcal{L}\{\partial f/\partial t\} = \int_0^\infty \frac{\partial f}{\partial t} \cdot \exp(-st)\, dt$$

$$= [f\exp(-st)]_0^\infty - \int_0^\infty -fs\exp(-st)\, dt$$

$$= -f(0) + s \int_0^\infty f\exp(-st)\, dt$$

$$= s\bar{f} - f(0)$$

(c)

$$\mathcal{L}\{\partial^2 f/\partial x^2\} = \int_0^\infty \frac{\partial^2 f}{\partial x^2} \cdot \exp(-st)\, dt$$

$$= \int_0^\infty \frac{\partial^2}{\partial x^2}(f\exp(-st))\, dt$$

$$= \frac{\partial^2}{\partial x^2} \int_0^\infty f\exp(-st)\, dt$$

$$= \frac{\partial^2 \bar{f}}{\partial x^2}$$

(ii) The diffusion equation states that:

$$\frac{\partial c}{\partial t} = D\frac{\partial^2 c}{\partial x^2}$$

We can take the Laplace transform of both sides:

$$s\bar{c} - c^* = D\frac{\partial^2 \bar{c}}{\partial x^2}$$

and rearrange:

$$\frac{\partial^2 \bar{c}}{\partial x^2} - \frac{s}{D}\bar{c} = -\frac{c^*}{D}$$

which is an inhomogeneous second-order differential equation with the solution:

$$\bar{c} = \frac{c^*}{s} + A(s)\exp\left(x\sqrt{\frac{s}{D}}\right)$$
$$+ B(s)\exp\left(-x\sqrt{\frac{s}{D}}\right)$$

We can Laplace transform the boundary conditions as well, so $\bar{c} = 0$ at $x \to \infty$ and so $A(s) = 0$. Then from the electrode boundary condition:

$$\bar{c}(0, s) = B(s) + \frac{c^*}{s} = 0$$

so $B(s) = -c^*/s$ and

$$\bar{c} = \frac{c^*}{s}\left(1 - \exp\left(-x\sqrt{\frac{s}{D}}\right)\right)$$

which on inverse transformation yields (using Eq. 3.1):

$$c = c^*\operatorname{erf}\left(\frac{x}{2\sqrt{Dt}}\right)$$

Since

$$I = nFAD\left.\frac{\partial c}{\partial x}\right|_{x=0}$$

we can derive the Cottrell equation

$$I = nFADc^*\frac{1}{2\sqrt{Dt}}\frac{2}{\sqrt{\pi}}\exp(0)$$

$$I = nFAc^*\sqrt{\frac{D}{\pi t}}$$

(iii) If we require the Nernst equation to hold, we may follow the same logic until the point of applying the electrode boundary condition to determine the unknown function $B(s)$. If all diffusion coefficients are equal, $c_{Ox} + c_{Red} = c^*$ everywhere. Then the Nernst equation for a one-electron reduction is:

$$\frac{F}{RT}(E - E_f^{\ominus}) = \ln\left(\frac{c_0}{c^* - c_0}\right)$$

which rearranges to

$$c_0 = \frac{c^*}{1 + \exp\left(-\frac{F}{RT}(E - E_f^\ominus)\right)}$$

Therefore in finding $B(s)$, recalling that the Laplace transform of a constant $k$ is $k/s$:

$$\bar{c}(0, s) = B(s) + \frac{c^*}{s} = \frac{c^*}{s} \frac{1}{1 + \exp\left(-\frac{F}{RT}(E - E_f^\ominus)\right)}$$

and so

$$B(s) = -\frac{c^*}{s} \frac{1}{1 + \exp\left(\frac{F}{RT}(E - E_f^\ominus)\right)}$$

Following forward:

$$\bar{c} = \frac{c^*}{s} \left( 1 - \frac{\exp\left(-x\sqrt{\frac{s}{D}}\right)}{1 + \exp\left(\frac{F}{RT}(E - E_f^\ominus)\right)} \right)$$

Hence, because both differentiation and inverse Laplace transformation are linear operations, we can note without further effort that

$$I = I_{Cottrell} \cdot \frac{1}{1 + \exp\left(\frac{F}{RT}(E - E_f^\ominus)\right)}$$

(iv) For 1% tolerance, we require

$$\frac{I}{I_{Cottrell}} = \frac{1}{1 + \exp\left(\frac{F}{RT}(E - E_f^\ominus)\right)} > 0.99$$

and hence

$$\ln(0.01/0.99) = -4.595 > F/RT(E - E_f^\ominus)$$

So, $E < E_f^\ominus - 4.595(RT/F)$, i.e. an overpotential of 118 mV from the formal reduction potential, at 298 K. If 0.1% tolerance is required, the same analysis reveals a required overpotential of 177 mV.

Of course, this all assumes that the electron transfer is fast. The analysis for slow electron transfer and unequal diffusion coefficients is equally feasible, but rather more involved; it is a recommended exercise in voltammetric theory which exceeds the scope of this book.

## 3.6  Diffusion and Root-Mean-Square Displacement

## Problem

For a molecule with diffusion coefficient $D$ travelling in one dimension $(x)$, the root-mean-square displacement in time $t$ is given:

$$\sqrt{\langle x^2 \rangle} = \sqrt{2Dt}$$

What is the corresponding expression $\sqrt{\langle r^2 \rangle}$ for a molecule moving in three dimensions $(x, y$ and $z)$?

## Solution

Since, by Pythagoras,

$$r^2 = x^2 + y^2 + z^2,$$

it follows that

$$\langle r^2 \rangle = \langle x^2 \rangle + \langle y^2 \rangle + \langle z^2 \rangle$$
$$= 2Dt + 2Dt + 2Dt$$
$$= 6Dt$$

Hence

$$\sqrt{\langle r^2 \rangle} = \sqrt{6Dt}$$

# 4

## Cyclic Voltammetry at Macroelectrodes

### 4.1 Cyclic Voltammetry: Electrochemically Reversible Voltammetry

#### Problem

Figure 4.1 shows the cyclic voltammetry associated with the reversible reduction of a solution phase species according to the equation below:

$$A + e^- \rightleftharpoons B$$

(i) Describe and explain the shape of the voltammogram.
(ii) How would the voltammogram differ if the species exhibited irreversible electron transfer kinetics?
(iii) What equation is used to parametrically describe the peak current of a voltammogram?

#### Solution

(i) The voltammogram shown in Fig. 4.1 is for a one-electron reduction of A to B where initially only species A is present in solution. The value of $E^{\ominus}_{f,A/B}$ for this reaction has been set as equal to 0 V. During the experiment the potential has been swept linearly from $+0.3\,\mathrm{V}$ to $-0.3\,\mathrm{V}$, after which the scan direction has been reversed (at $-0.3\,\mathrm{V}$) and the potential has been swept linearly back to $+0.3\,\mathrm{V}$. The arrows shown in Fig. 4.1 indicate the scan direction.

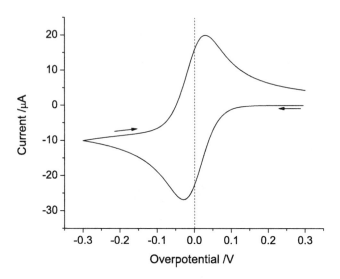

**Fig. 4.1** Cyclic voltammogram for the reduction of A to B. Arrows indicate the scan direction; the start potential is +0.3 V.

For the forward scan, three distinct regions are observed:

- Between +0.3 V and $\approx +0.15$ V, no current is passed as there is insufficient driving force (overpotential) for A to be reduced to B.
- As the potential is further decreased the rate of reduction of A to B increases. The measured current increases approximately exponentially. Since the electron transfer kinetics are 'fast' in nature (i.e. the species is electrochemically reversible) the concentrations of species A and B at the electrode surface obey the Nernst equation (Eq. 4.1).

$$\frac{[A]_0}{[B]_0} = \exp(\theta) \tag{4.1}$$

where

$$\theta = \frac{F}{RT}\left[E - E^{\ominus}_{f,A/B}\right]$$

Accordingly, at zero overpotential, as marked on Fig. 4.1 by the dashed line, the concentrations of species A and B at the electrode surface are equal.
- As the potential is further decreased, the absolute current goes through a maximum (at −0.03 V) due to depletion of A at the electrode surface. At more negative potentials the measured current is now limited by the diffusion of species A to the electrode surface.

On reversing the potential, the concentrations of species A and B continue to obey the Nernst equation and hence scanning in the positive direction, a peak is observed at $+0.03$ V. This is associated with the reoxidation to species B to A and the resulting depletion of B at the electrode surface.

The interpretation of a voltammogram is usually done on the basis of the measurement of the peak current for the forward scan and the voltammetric peak-to-peak separation of the forward and reverse scans. For a reversible one-electron redox system the peak-to-peak separation should ideally be 57 mV (at 25°C), but due to the finite size of the potential window it is usually slightly elevated to 59–60 mV even for reversible kinetics.

(ii) As the standard electrochemical rate constant is decreased, the most notable change in the observed voltammetry is the increase in the peak-to-peak separation. Further, the peak current will also be less than that found for a fully reversible redox system. This latter point is explored in Problem 4.8.

(iii) The peak current of a voltammogram is described parametrically by the Randles–Ševčík equation. The form given below is for a reversible $n$ electron redox species:

$$I_{pf} = (2.69 \times 10^5) n^{\frac{3}{2}} A c^* D^{\frac{1}{2}} v^{\frac{1}{2}} \qquad (4.2)$$

where $I_{pf}$ (A) is the peak current associated with the forward scan, $n$ is the number of electrons transferred, $A$ (cm$^2$) is the geometric area of the electrode, $c^*$ (mol cm$^{-3}$) is the bulk concentration of species A, $D$ (cm$^2$ s$^{-1}$) is the diffusion coefficient of species A and $v$ (V s$^{-1}$) is the experimental scan rate.

Note that care must be taken when using this equation to ensure that all values are expressed in consistent units (lengths in either m or cm), specifically noting that in the above form concentration is expressed in terms of moles per cubic centimetre. As a further caveat we emphasise that use of this equation is *not* suitable for analysing the peak in the reverse scan.

## 4.2 Cyclic Voltammetry: Electrochemically Irreversible Voltammetry

### Problem

Figure 4.2 depicts a cyclic voltammogram for the one-electron irreversible reduction of species A to B, where the electrochemical rate constant ($k^0$) equals $10^{-6}$ cm s$^{-1}$ and the transfer coefficient ($\alpha$) is 0.5. Problem 4.1 discussed the salient features of a reversible cyclic voltammogram; with reference to this, explain the major differences between the two cases. Specifically refer to reasons:

- why the peak-to-peak separation is greater for the irreversible case.
- why the reverse peak current for the irreversible case is substantially less than the forward peak current.

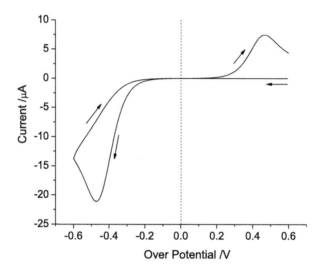

**Fig. 4.2** Cyclic voltammogram for the irreversible reduction of A to B. Arrows indicate the scan direction. The current has been plotted against overpotential $\left(E - E^{\ominus}_{f,A/B}\right)$.

## Solution

As with the reversible case, the forward wave may be described by three distinct regions:

- At low overpotential, zero current is passed as there is insufficient driving force (overpotential) for A to be reduced to B.
- As the potential is further decreased the rate of reduction increases in accordance with the Butler–Volmer equation:

$$I = -F\,Ak^0 \left( \exp\left[ \frac{-\alpha F}{RT}(E - E^{\ominus}_f) \right] [A]_0 - \exp\left[ \frac{(1-\alpha)F}{RT}(E - E^{\ominus}_f) \right] [B]_0 \right)$$

$$(4.3)$$

where $I$ is the current, $F$ is the Faraday constant, $\alpha$ is the transfer coefficient ($\approx 0.5$), $A$ is the area of the electrode, $k^0$ is the *standard electrochemical rate constant* and $[i]_0$ is the concentration of species $i$ at the electrode.

- As the potential decreases further, the current goes through a maximum (at $-0.47$ V) due to depletion of A at the electrode surface. At this point and at more negative potentials, the current is limited by the flux of A to the electrode surface, due to diffusion.

Because the redox couple exhibits a low $k^0$, the concentrations of species A and B are no longer directly related by the Nernst equation. Rather, as stated above, the current may be described in terms of the Butler–Volmer equation. Hence it is necessary for a relatively large overpotential to be applied to the electrode in order for the reduction to occur at a measurable rate. It is the need for this overpotential to drive electron transfer, and hence the delay in depletion of the electroactive species at the surface, which leads to the large peak-to-peak separation.

Further, as a result of this large peak-to-peak separation, the reduced species B which is produced during the forward scan is better able to diffuse away from the electrode without being prone to reoxidation – the diffusion occurs because the concentration of B is zero in the bulk solution. As such, on returning to the overpotential required for the reoxidation of B to A, a significant decrease in the concentration of B adjacent to the electrode has occurred. This leads to the magnitude of the reverse peak being less than that of the forward.

## 4.3  Reversible vs Irreversible Voltammetry

## Problem

When analysing a voltammogram it is important to consider whether the electron transfer rate is reversible or irreversible in nature. Discuss the main features of a voltammogram that can be readily used to qualitatively decide whether a voltammogram indicates 'reversible' or 'irreversible' electron transfer.

## Solution

Consideration of three main features allows an experimentalist to assess if a voltammogram is exhibiting reversible electron transfer or not. These are as follows:

(i) *The peak-to-peak separation*: for a one-electron transfer at 25°C the peak-to-peak separation should approach 57 mV in the fully reversible limit. A value greater than 60 mV suggests that the electron transfer is either quasi-irreversible or irreversible.

   Further, the peak-to-peak separation for a voltammogram will vary as a function of scan rate for non-reversible electron transfer. The separation will be much larger for the irreversible case.

(ii) *The peak current*: in both cases the peak current varies with the square root of scan rate, but the constant of proportionality of this variation differs. This is clearly exemplified through consideration of the forms of the Randles–Ševčík equation.

For a reversible one-electron process:

$$|I_{pf}| = (2.69 \times 10^5) A c^* D^{\frac{1}{2}} v^{\frac{1}{2}} \tag{4.4}$$

and for an irreversible one electron-process:

$$|I_{pf}| = (2.99 \times 10^5)\alpha^{\frac{1}{2}}Ac^*D^{\frac{1}{2}}v^{\frac{1}{2}} \qquad (4.5)$$

where $I_{pf}$ (A) is the peak current associated with the forward scan, $n$ is the number of electrons transferred, $A$ (cm$^2$) is the geometric area of the electrode, $c^*$ (mol cm$^{-3}$) is the bulk concentration of the redox species, $D$ (cm$^2$ s$^{-1}$) is the diffusion coefficient of the redox species, $v$ (V s$^{-1}$) is the experimental scan rate and $\alpha$ is the transfer coefficient.

(iii) *The waveshape*: the forward peak waveshape for an irreversible redox system differs distinctly from that of a reversible voltammogram. The waveshape for a reversible redox couple appears 'sharper' than that for an irreversible redox couple. This may be quantified through considering the difference in potential between the peak current and the half-peak current.

Tafel analysis of the voltammetric wave will provide the same information. From Problems 2.3 and 2.4, we know that for a one-electron irreversible wave, Tafel analysis yields a line of gradient equal to $\alpha F/RT$, whereas for a reversible one-electron wave the gradient is equal to $F/RT$, such that the exponential portion (Tafel region) of the voltammogram is steeper in the reversible case.

## 4.4 Voltammetric Diagnostics

### Problem

The peak current of a voltammogram is a readily measurable quantity for diagnostic purposes. Specifically, the Randles–Ševčík equations as defined in Problem 4.3 are used to parametrically define this measurement. Explain why in the following cases these equations are *not* applicable, and 'peak current' is a misleading parameter:

(i) The analysis of the back peak of a voltammogram.
(ii) The analysis of the forward peak on the second scan of a voltammogram.

### Solution

The Randles–Ševčík equations are derived from the assumption that the concentration of the species at the electrode surface at the beginning of the linear voltammetric scan is equal to that in bulk. This is correct for a forward scan, but is not correct for the two cases outlined above because electron transfer has already taken place. In this case, a diffusion layer has already been developed and so the system is perturbed.

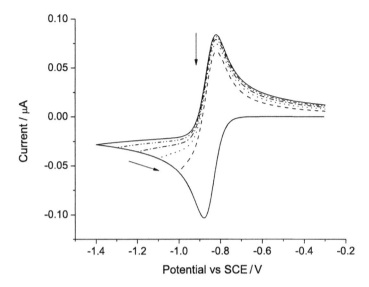

**Fig. 4.3** A simulated voltammogram of a one-electron reduction, highlighting the influence of the switching potential upon the height of the back peak.

The observed voltammetry is then a *superposition* of currents from different Faradaic processes and so care must be taken in discussing 'peak currents'. These are not usually meaningful when compared to a baseline of zero current. Hence, the use of the Randles–Ševčík equations to analyse peak currents in either of the above cases will result in significant errors. As a further point, it should be noted that the size and potential of the reverse peak vary as a function of the switching potential, as highlighted in Fig. 4.3 (unless, for a reversible redox couple, the switching potential is beyond a certain threshold).

## 4.5 Voltammetry and Scan Rate Effects

### Problem

Cyclic voltammetry (as shown in Fig. 4.4 where the current has been normalised with respect to the square root of the scan rate) was recorded for a compound A (with $D = 10^{-5}$ cm$^2$ s$^{-1}$) at a large electrode, at scan rates ranging from 2 mV s$^{-1}$ to 2 V s$^{-1}$. Account for the observed change in voltammetry. Is it possible to infer any kinetic information from these data?

### Solution

At the lowest scan rates, the voltammetry is unaltered with scan rate, except for the current scaling by $v^{\frac{1}{2}}$. The scan rate is sufficiently slow that the kinetics of

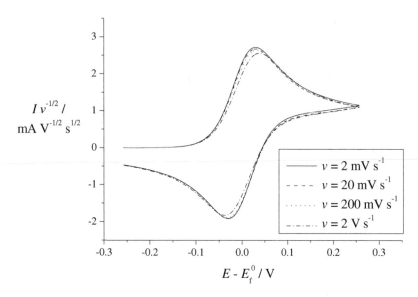

**Fig. 4.4** Cyclic voltammetry at four scan rates for the one-electron oxidation of 1 mM of species A $(D = 10^{-5}\,\text{cm}^2\,\text{s}^{-1})$ on a 1 cm radius disc electrode.

the electrolysis process are not outrun, and so Nernstian (reversible) behaviour is observed. At faster scan rates, an increase in peak-to-peak separation is noted because the scan is now fast enough that the electrode kinetics become rate-limiting.

The Matsuda–Ayabe parameter, $\Lambda$, quantifies kinetic reversibility for an electrochemical process:

$$\Lambda = \frac{k^0}{\left(\frac{FDv}{RT}\right)^{\frac{1}{2}}}$$

At the point of transition between reversible voltammetry and quasi-reversibility, $\Lambda \simeq 15$. Therefore for the lower two scan rates, we can infer that $\Lambda \geq 15$. For the faster two scan rates, the behaviour is quasi-reversible so we expect that the Matsuda–Ayabe parameter is $<15$.

Given that the transition occurs at approximately $50\,\text{mV s}^{-1}$, by substitution:

$$\frac{k^0}{\left(\frac{F}{RT}(10^{-5} \times 0.05)\right)^{\frac{1}{2}}} \approx 15$$

$$k^0 \approx 0.07\,\text{cm s}^{-1}$$

This is a necessarily approximate result and a more detailed study would be necessary to determine $k^0$ precisely; in particular, numerical simulation might be employed. However, the result is accurate to within an order of magnitude.

## 4.6 Ferrocene Voltammetry

## Problem

The peak current for a 1 mM solution of ferrocene in acetonitrile was measured as a function of scan rate on an electrode with an area of $0.1\,cm^2$. The table below gives the experimentally measured values. What is the diffusion coefficient for this species?

| Scan rate $v$/mV s$^{-1}$ | Peak current $I_{pf}$/$\mu$A |
|---|---|
| 25 | 21.3 |
| 50 | 30.0 |
| 100 | 42.5 |
| 200 | 60.2 |
| 400 | 85.1 |

## Solution

Ferrocene is a fully reversible redox species at these scan rates. Hence the following form of the Randles–Ševčík equation must be used to assess the diffusion coefficient:

$$I_{pf} = (2.69 \times 10^5) n^{\frac{3}{2}} A D^{\frac{1}{2}} c^* v^{\frac{1}{2}}$$

where $I_{pf}$ is the peak current in amperes, $A$ is the area of the electrode surface in cm$^2$, $D$ is the diffusion coefficient of the species in cm$^2$ s$^{-1}$, $c^*$ is the bulk concentration of the species in mol cm$^{-3}$ and $v$ is the scan rate in V s$^{-1}$.

A plot of peak current versus $v^{\frac{1}{2}}$ yields a straight line with a gradient proportional to the square root of the diffusion coefficient of the species. From a plot of the data above, the gradient is $134.6\,\mu A\,V^{-\frac{1}{2}}\,s^{\frac{1}{2}}$. The diffusion coefficient of ferrocene is hence found to be $2.5 \times 10^{-5}\,cm^2\,s^{-1}$.

## 4.7 Ferrocene Voltammetry

## Problem

Discuss critically the statement that the ferrocene/ferrocenium redox couple is fully electrochemically reversible in acetonitrile solution at a platinum electrode.

## Solution

Electrochemical reversibility requires that the electron transfer kinetics are fast relative to the prevailing rates of mass transport. As such the notion of 'reversiblity' is not simply a property of the redox couple but also of the conditions, especially mass transport, under which it is measured.

N.V. Rees *et al.* [*J. Electroanal. Chem.* **580** (2005) 78] showed that for Fc/Fc$^+$ in CH$_3$CN at 25°C, $k^0 \approx 1 \text{ cm s}^{-1}$ at a platinum electrode. The mass transport coefficient to an electrode of radius $r_e$ is:

$$k_m = \frac{D}{r_e}$$

so since $D \simeq 2 \times 10^{-5} \text{ cm s}^{-1}$, electrodes of the radius $r_e \leq 1\mu\text{m}$ will show quasi-reversible behaviour.

## 4.8 Features of Cyclic Voltammograms

### Problem

Account for the following features of cyclic voltammograms:

(i) Why is it expected that peak current is proportional to the square root of scan rate?

(ii) Why is the peak current less in the limit of slow electrode kinetics?

(iii) Why does altering the diffusion coefficient of the product ($D_B$) for a given diffusion coefficient of the reactant ($D_A$) affect the forward peak potential but not the forward peak current?

(iv) Why, in practice, does a plot of $I_{pf}$ vs. $c^*$ often not go through the origin?

### Solution

(i) The peak current occurs when the electroactive species becomes depleted at the electrode surface. The thickness of the diffusion layer is related to the time in which depletion has taken place by the Einstein equation: $x_D \propto \sqrt{Dt}$. So $x_D \propto v^{-1/2}$.

The current is related to the flux across the electrode surface by Faraday's laws; for a fixed concentration drop – bulk to zero – this flux is then inversely proportional to diffusion layer thickness. Hence $I \propto x_D^{-1} \propto v^{1/2}$. In effect, the diffusion layer becomes narrower in proportion to the square root of scan rate, and so the flux is elevated in the same proportion.

(ii) The peak current represents the rate of electron transfer when mass transport becomes rate-determining. In the absence of mass transport effects, a reversible reaction will increase its rate in proportion to the thermodynamic driving force, i.e. $\propto \exp(nvFt/RT)$. For slow electrode kinetics, by comparison, the rate varies only as quickly as the activation energy, $\Delta G^\ddagger$, can be lowered. According to the Butler–Volmer theory of electrode kinetics, this takes place as $\exp(v\alpha Ft/RT)$ where $0 \leq \alpha \leq n$.

Therefore, the rate of electron transfer for quasi-reversible kinetics does not increase with the swept potential as quickly as if $\Delta G^\ddagger$ is sufficiently low as to be irrelevant. Consequently, the rate of electron transfer at the point of depletion of material at the electrode surface is lower than in the reversible case, and correspondingly the peak current in a quasi-reversible case is less.

(iii) The forward peak current arises at the point where the supply of material to the electrode becomes rate-determining, in place of the electrode kinetics. The rate at which this supply takes place depends only on the diffusion coefficient of the reactant, $D_A$, and so it is unaffected by $D_B$. The diffusion coefficients also do not affect the electrode kinetics which depend on the electrochemical couple and the interface, as quantified by $k^0$ and $\alpha$.

By contrast, the potential at which this limit occurs can be altered because a faster $D_B$ causes more efficient dispersion of the product from the electrode. Then further reaction is driven, i.e. the equilibrium is shifted according to Le Chatelier's principle. So, a correspondingly faster reaction occurs at lower overpotentials. Therefore $D_B$ does affect the forward peak potential in an *apparently* electrocatalytic manner.

(iv) In a real electrochemical setup, it is common to encounter a capacitive current which arises due to the response of the electrical double layer to the swept potential (see also Problem 10.6). As the potential changes, the charge on the electrode also changes and so ions in solution will move in response to this charge by virtue of their attraction or repulsion from the surface, even if they are not oxidised or reduced. This induces a measurable current, recorded by the potentiostat, which is not due to electrolysis.

Only the Faradaic current resulting from the electrolysis process varies linearly with $c^*$; the capacitive current generally tends to depend on the total salt concentration which is dominated by the supporting electrolyte. Therefore the overall peak current, $I_{pf}$, is a linear function of $c^*$ but is not zero in the absence of a redox-active species since a capacitive current is still present. When interpreting experimental data it is important to carefully choose a baseline from which the Faradaic current may be measured and hence allow analytically useful data to be obtained.

## 4.9 Derivation of the Randles–Ševčík Equation

### Problem

Unlike chronoamperometry, exact solutions do not exist for cyclic voltammetry problems. Mathematical methods can lead to useful results, however.

(i) How is a cyclic voltammetry problem described mathematically?

(ii) Show that the surface concentrations at any point during a cyclic voltammogram can be expressed as a function of the current. Assume species A to undergo an $n$-electron reduction to species B.

(iii) Hence show that if the Nernst equation holds at the surface, the current is proportional to some potential-dependent function $\chi(E)$.

(iv) In the textbook form of the Randles–Ševčík equation, the cathodic peak current of a reversible cyclic voltammogram is $I_{pf} = -2.69 \times 10^5 \, n^{\frac{3}{2}} A D_A^{\frac{1}{2}} c^* v^{\frac{1}{2}}$, with all symbols defined as for Eq. 4.2. Determine the maximum value of the function $\chi(E)$.

## Solution

(i) A cyclic voltammetry problem is treated mathematically by solving the diffusion equation (Fick's second law) for each chemical species present. For a macroelectrode, we can ignore the electrode edge and solve in one-dimensional planar space, just as in deriving the Cottrell equation, which is generally simpler than a radially symmetric or multi-dimensional solution. This assumes that sufficient electrolyte has been added to screen electric fields, and that the solution is not stirred or subject to natural convection.

Two boundary conditions are required for each species: one in bulk and one at the electrode. Conventionally the concentrations are set to their initial values in bulk, and at the electrode either the Nernst equation or Butler–Volmer equation is applied to describe the electrode kinetics. These equations have the general form $f(c_{i,0}, E) = 0$, where the applied potential $E$ is a linear function of time. Conservation of mass also requires that the fluxes of reactant A and product B are equal and opposite at the electrode surface.

(ii) The transport equations for species A and B are:

$$\frac{\partial c_A}{\partial t} = D_A \frac{\partial^2 c_A}{\partial x^2}$$

$$\frac{\partial c_B}{\partial t} = D_B \frac{\partial^2 c_B}{\partial x^2}$$

with $c_A = c^*$ and $c_B = 0$ at $t = 0$ and $x = \infty$. Additionally, from conservation of mass we know that at $x = 0$:

$$D_A \frac{\partial c_A}{\partial x} = -D_B \frac{\partial c_B}{\partial x}$$

The transformations $c_i' = c_i/c^*$, $D_i' = D_i/D_A$ and $\tau = D_A t$ are immediately useful as they relegate two of the parameters to scaling factors. Then if we Laplace transform all the equations from the coordinate $\tau$ to a Laplace coordinate $s$:

$$s\bar{c}_A - 1 = \frac{\partial^2 \bar{c}_A}{\partial x^2}$$

$$s\bar{c}_B = \frac{\partial^2 \bar{c}_B}{\partial x^2}$$

in which we have denoted $\bar{c} = \mathcal{L}\{c'\}$. With application of the outer boundary conditions these readily solve to:

$$\bar{c}_A = \frac{1}{s} + \lambda_A(s)\exp\left(-x\sqrt{s}\right)$$

$$\bar{c}_B = \lambda_B(s)\exp\left(-x\sqrt{\frac{s}{D'_B}}\right)$$

with $\lambda_i(s)$ being as yet unknown functions.

Differentiating the above and substituting in $x = 0$, and then considering the boundary condition for conservation of mass, it follows that:

$$\lambda_A\sqrt{s} = -D'_B\lambda_B\sqrt{\frac{s}{D'_B}}$$

and therefore

$$\lambda_A = -\sqrt{D'_B}\cdot\lambda_B$$

On substitution of $x = 0$ into the expressions for $\bar{c}_i$, we note that this implies:

$$c'_{A,0} = 1 - \sqrt{D'_B}\cdot c'_{B,0}$$

and so the two surface concentrations are related at all times.

We also recognise that for a reduction

$$j(t) = \frac{-I(\tau)}{nFAD_A c^*} = \left.\frac{\partial c_A}{\partial x}\right|_{x=0}$$

and therefore

$$\mathcal{L}\{j(\tau)\} = \sqrt{s}\cdot\lambda_A(s) = \sqrt{s}\,\mathcal{L}\{c'_{A,0} - 1\}$$

so

$$c'_{A,0} = 1 + \mathcal{L}^{-1}\left\{\frac{\mathcal{L}\{j(\tau)\}}{\sqrt{s}}\right\}$$

Here we need the convolution theorem which identifies the inverse Laplace transform of a product as being a convolution:

$$\mathcal{L}^{-1}\{\bar{f}(s)\bar{g}(s)\} = \int_0^\tau f(\tau - \tau')g(\tau')d\tau'$$

and so for the above example simply:

$$c_{A,0} = 1 + \int_0^\tau \frac{j(\tau')}{\sqrt{\tau - \tau'}}d\tau'$$

Together with the above, both $c'_{i,0}$ can now be written as (integral) functions of the current $I(\tau)$.

(iii)  From the Nernst equation, we know that for a reduction:

$$E = E_f^\ominus + \frac{RT}{nF} \ln \frac{c_{A,0}}{c_{B,0}}$$

and so

$$c_{A,0} = c_{B,0} \exp\left( \frac{nF}{RT}(E - E_f^\ominus) \right)$$

$$1 + \int_0^\tau \frac{j(\tau')}{\sqrt{\tau - \tau'}} d\tau' = \frac{-1}{\sqrt{D_B'}} \left( \int_0^\tau \frac{j(\tau')}{\sqrt{\tau - \tau'}} d\tau' \right) \times \exp\left( \frac{nF}{RT}(E - E_f^\ominus) \right)$$

and rearranging

$$\int_0^\tau \frac{j(\tau')}{\sqrt{\tau - \tau'}} d\tau' = \frac{-1}{1 + \frac{1}{\sqrt{D_B'}} \exp\left( \frac{nF}{RT}(E - E_f^\ominus) \right)}$$

We note that the expression on the right varies in time only as a function of

$$\frac{nF}{RT} E = \frac{nF}{RT} E_i + \frac{nFvt}{RT}$$

$$= \ln a + \sigma\tau$$

where $a$ is a constant and $\sigma = \frac{nF}{RT} \frac{v}{D_A}$. The integral on the left may be written in terms of $u = \sigma\tau$:

$$\int_0^\tau \frac{j(\tau')}{\sqrt{\tau - \tau'}} d\tau' = \int_0^u \frac{j(u')}{\sqrt{\frac{u - u'}{\sigma}}} \frac{du}{\sigma}$$

$$= \int_0^u \frac{\frac{j(u')}{\sqrt{\sigma}}}{\sqrt{u - u'}} du$$

$$= \frac{-1}{1 + \frac{a}{\sqrt{D_B'}} \exp\left( \sigma\tau \right)} \tag{4.6}$$

Therefore $j(\sigma\tau)/\sqrt{\sigma}$ has some solution $\chi(E)$ defined by the above integral equation, and so:

$$I(E) = -nFAD_A C^* \sigma^{1/2} \chi(E)$$

$$= -n^{3/2} FAD_A^{1/2} c^* \sqrt{\frac{Fv}{RT}} \chi(E)$$

where $\chi(E)$ can be determined by numerical solution of its defining integral equation (Eq. 4.6).

(iv)  Our result from part (iii) evidently resembles the Randles–Ševčík equation closely. It follows that where $\chi(E)$ is maximal, $\sqrt{F^3/RT}\chi(E) = 2.69 \times 10^5$. Dividing by $\sqrt{F^3/RT}$ we find that $\chi(E) \simeq 0.447$ at its maximum.

## 4.10  Reversible Two-Electron Transfer

## Problem

For the following mechanism

$$A + e^- \rightleftharpoons B \quad E^{\ominus}_{f,A/B}$$

$$B + e^- \rightleftharpoons C \quad E^{\ominus}_{f,B/C}$$

sketch and explain the observed voltammetric response when

(i) $E^{\ominus}_{f,B/C} \gg E^{\ominus}_{f,A/B}$
(ii) $E^{\ominus}_{f,B/C} \ll E^{\ominus}_{f,A/B}$

Assume fast electron transfer kinetics.

## Solution

Figure 4.5 shows the voltammetric response for the two-electron reductions of A to C. In both cases $E^{\ominus}_{f,A/B} = 0.0\,\text{V}$.

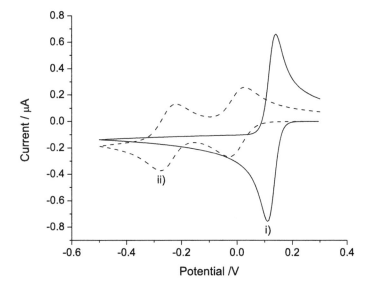

**Fig. 4.5** Simulated cyclic voltammograms for a two-electron reduction: where (i) the second electron transfer occurs at a more positive potential than the first $\left(E^{\ominus}_{f,B/C} \gg E^{\ominus}_{f,A/B}\right)$; and (ii) the second electron transfer occurs at a more negative potential than the first $\left(E^{\ominus}_{f,B/C} \ll E^{\ominus}_{f,A/B}\right)$.

In case (i), where the second electron transfer occurs at a more positive potential than the first ($E_{f,B/C}^{\ominus} = +0.25\,V$), only one voltammetric wave is observed. This corresponds to the overall two-electron reduction of A to C.

In case (ii), the second electron transfer occurs at a more negative potential than the first ($E_{f,B/C}^{\ominus} = -0.25\,V$). Here, two separate voltammetric signals are observed. The first wave corresponds to the reduction of A to B and the second corresponds to the reduction of B to C.

It is interesting to note that in case (i) the reduction of A to C occurs at a potential greater than $E_{f,A/B}^{\ominus}$, i.e. above 0.0 V. This is due to the presence of the second electrochemical step, which is more favourable and so drives the first electrochemical step by consumption of the product B. The voltammetry is under thermodynamic control so that the apparent value of $E_f^{\ominus}$ for the overall process is given by $\left(E_{f,A/B}^{\ominus} + E_{f,B/C}^{\ominus}\right)/2$.

## 4.11 The Influence of pH on Cyclic Voltammetry

## Problem

For the following reaction schemes, describe how the voltammetric response for the redox reactions will vary as a function of pH over the full aqueous pH range (0–14). Assume that the rate of electron transfer is fast, such that the electrode kinetics are fully reversible, and that the protonation/deprotonation steps are so fast as to be equilibrated throughout. Note that the chemical steps have been written as deprotonations so that the equilibrium constants are readily related to the $pK_a$ of the species.

(i)

$$A + e^- \rightleftharpoons A^- \quad E_{f,A/A^-}^{\ominus} = 0.0\,V$$

$$AH \rightleftharpoons A^- + H^+ \quad (K_{a2})$$

where

$$K_{a2} = \frac{[A^-][H^+]}{[AH]} = 10^{-11}$$

(ii)

$$AH^+ \rightleftharpoons A + H^+ \quad (K_{a1})$$

$$AH^+ + e^- \rightleftharpoons AH \quad E_{AH^+/AH}^{\ominus} = +0.473\,V$$

where

$$K_{a1} = \frac{[A][H^+]}{[AH^+]} = 10^{-3}$$

## Solution

Given that the electron transfer is assumed to be fully reversible, the midpoint potential will be equal to the formal potential for the reduction (assuming that all diffusion coefficients are equal). Hence the voltammogram will shift with the formal potential for the reduction as a function of pH. Through following an analogous methodology to that in Problem 1.26, we can derive the following equations to describe the shifts in the formal potentials.

(i)

$$E_{f,A/A^-}^{\ominus} = E_{A/A^-}^{\ominus} - \frac{RT}{F} \ln \frac{[AH]_{tot}}{[A]} + \frac{RT}{F} \ln \left(1 + \frac{[H^+]}{K_{a2}}\right)$$

(ii)

$$E_{f,AH^+/AH}^{\ominus} = E_{AH^+/AH}^{\ominus} - \frac{RT}{F} \ln \frac{[AH]}{[A]_{tot}} - \frac{RT}{F} \ln \left(1 + \frac{K_{a1}}{[H^+]}\right)$$

where

$$[AH]_{tot} = [A^-] + [AH]$$
$$[A]_{tot} = [A] + [AH^+]$$

Figure 4.6 plots how these formal potentials vary as a function of pH.

For system (i), below pH 11 the reaction involves the transfer of one proton and one electron, as the $A^-$ formed is protonated to give AH. Hence the voltammetric peaks will shift with approximately 59 mV per pH (at 25°C). Above pH 11 the reaction does not involve proton transfer as $A^-$ is not protonated at these pH values. Consequently, the peak position is invariant with pH.

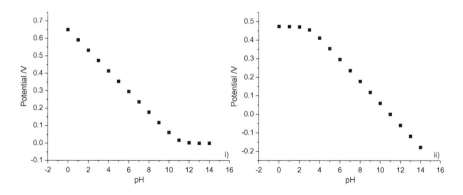

**Fig. 4.6** Plots of the midpoint potential versus pH for the reactions (i) and (ii).

For system (ii), below pH 3 species A is protonated in solution. Hence the voltammogram shows the reduction of $AH^+$ to AH. No protons are transferred during the electrochemical process and hence the peak does not shift with pH. Above pH 3, species A is present in solution and hence for it to be reduced, it must first be protonated. Consequently, the voltammetric peak will shift with approximately 59 mV per pH (at 25°C).

## 4.12 The Scheme of Squares

## Problem

The scheme of squares is used to describe mechanistic pathways involving electron and proton transfers, and was first proposed by J. Jacq [*J. Electroanal. Chem.* **29** (1971) 149]. The scheme is based upon the assumption that the reactions occur in a stepwise manner. Figure 4.7 depicts a simple one-proton one-electron scheme. In the following, assume $E_1 = 0.0$ V, $E_2 = +0.473$ V, $pK_{a1} = 3$ and $pK_{a2} = 11$.

Although the values given above are only hypothetical, a large number of organic and inorganic electrochemical processes are found to follow a square scheme.

(i) Assuming reversible electron transfer, sketch how the voltammetric midpoint potential will shift as a function of pH over the full aqueous pH range (0–14).

(ii) In a number of systems it has been suggested that the proton and electron are not transferred in a stepwise mechanism, but rather that both transfer in a concerted manner. How may it be possible to verify experimentally whether or not the protons and electrons are being transferred in a concerted or stepwise manner, for a given electrochemical reaction?

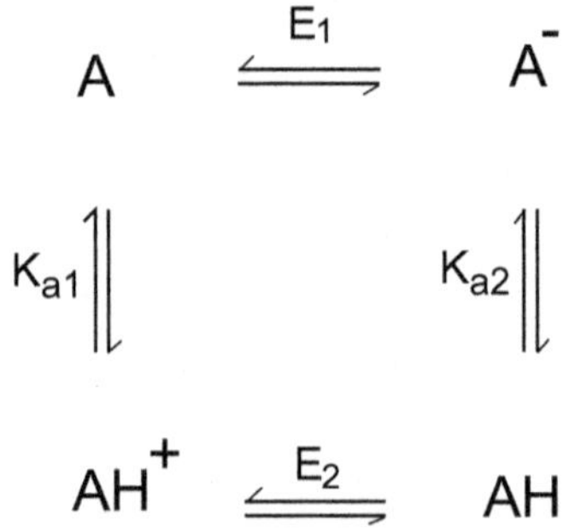

Fig. 4.7  A one-electron one-proton scheme of squares.

## Solution

(i) The scheme of squares outlined in Fig. 4.7 may be viewed as being a result of the combination of the two reaction mechanisms outlined in Problem 4.11, parts (i) and (ii). At pH below 3, the voltammetric response will correspond to the one-electron reduction of $AH^+$ to AH. Hence no protons are transferred and the peak position will be insensitive to pH.

At pH values above three but below 11, the overall reaction will be the one-proton one-electron reduction of species A to AH. Consequently, the voltammetric signal will be found to shift in potential with approximately 59 mV per pH (at 25°C), irrespective of whether electron transfer or proton transfer occurs first.

Above pH 11, the reaction corresponds to the reduction of species A to $A^-$ without proton transfer. Accordingly the peak position will not vary with pH. This result is shown within Fig. 4.8.

(ii) If the electron and proton are transferred in a concerted manner, then the rate of electron transfer should exhibit a *kinetic* isotope effect, as evidenced by a change in the peak-to-peak separation. Accordingly, if the proton–electron redox reaction is performed in a solution of $D_2O$, the *rate* of electron transfer should be slower. This concerted mechanism is thought to be highly important

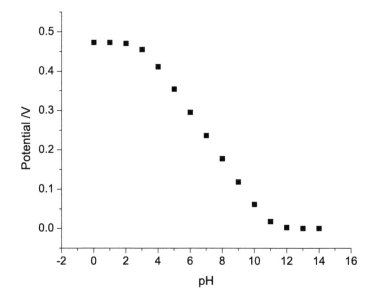

**Fig. 4.8** The variation of the midpoint potential for the scheme of squares, as outlined in Fig. 4.7.

in a number of biological redox processes. For a recent review on the area see C. Costentin *et al.* [*Acc. Chem. Res.* **43** (2010) 1019].

## 4.13 The EE-Comproportionation Mechanism

### Problem

Two-electron transfers (EE) in which the second electron transfer is less favourable than the first are very commonly encountered in voltammetry. If we denote the species involved as A, B and C, respectively, then the comproportionation reaction is favourable:

$$A + e^- \rightleftharpoons B$$

$$B + e^- \rightleftharpoons C$$

$$A + C \overset{K_{comp}}{\rightleftharpoons} 2B$$

(i) Use Hess's law to determine how the equilibrium constant $K_{comp}$ is related to the reduction potentials for the heterogeneous reactions. If $E^{\ominus}_{f,A/B} \gg E^{\ominus}_{f,B/C}$, will this reaction be favourable?

(ii) Demonstrate that if $D_A = D_B = D_C$ and the electrochemical reaction is reversible, the current is blind to the kinetics of the comproportionation reaction, irrespective of the voltammetric waveform.

(iii) Various researchers have used diverse methods to determine the rate of comproportionation for mechanisms of this type. Examine the following articles and discuss how the problem is tackled in each case:

    (a) C.P. Andrieux *et al.*, *J. Electroanal. Chem.* **28** (1970) 339.
    (b) Z. Rongfeng *et al.*, *J. Electroanal. Chem.* **385** (1995) 201.
    (c) S.R. Belding *et al.*, *Angew. Chem. Intl. Ed.* **49** (2010) 9242.

### Solution

(i) The comproportionation is a combination of the reduction of A to B and the oxidation of C to B. Therefore:

$$\Delta G^{\ominus}_{comp} = -FE^{\ominus}_{comp} = -F\left(E^{\ominus}_{f,A/B} - E^{\ominus}_{f,B/C}\right)$$

and using the common relation of Gibbs energy change to the equilibrium coefficient:

$$K_{comp} = \exp\left(\frac{-\Delta G^{\ominus}_{comp}}{RT}\right) = \exp\left(\frac{F}{RT}\left(E^{\ominus}_{f,A/B} - E^{\ominus}_{f,B/C}\right)\right)$$

If $E^{\ominus}_{f,A/B} \gg E^{\ominus}_{f,B/C}$, it is clear that $K_{comp} \gg 1$ and so the comproportionation is thermodynamically favoured.

(ii) The mathematical formulation of the problem for species A, B and C will be:

$$\frac{\partial c_A}{\partial t} = D_A \frac{\partial^2 c_A}{\partial x^2} - k_{comp} c_A c_C + k_{disp} c_B^2$$

$$\frac{\partial c_B}{\partial t} = D_B \frac{\partial^2 c_B}{\partial x^2} + 2 k_{comp} c_A c_C - 2 k_{disp} c_B^2$$

$$\frac{\partial c_C}{\partial t} = D_C \frac{\partial^2 c_C}{\partial x^2} - k_{comp} c_A c_C + k_{disp} c_B^2$$

where $k_{comp}$ and $k_{disp}$ are the second-order rate constants for the homogeneous reaction, and so $K_{comp} = k_{comp}/k_{disp}$.

If the reaction is reversible, the Nernst equation relates the concentrations of A, B and C at the electrode surface, and these are also related by conservation of mass. Hence at $x = 0$:

$$c_A = \exp\left(\frac{F}{RT}\left(E - E^{\ominus}_{f,A/B}\right)\right) c_B$$

$$c_B = \exp\left(\frac{F}{RT}\left(E - E^{\ominus}_{f,B/C}\right)\right) c_C$$

$$D_A \frac{\partial c_A}{\partial x} + D_B \frac{\partial c_B}{\partial x} + D_C \frac{\partial c_C}{\partial x} = 0$$

If $D_A = D_B = D_C = D$, we can make the substitutions $\sigma = c_A + c_B + c_C$ and $u = 2c_A + c_B$, thus cancelling the homogeneous reaction terms from the transport equations:

$$\frac{\partial \sigma}{\partial t} = D \frac{\partial^2 \sigma}{\partial x^2}$$

with $\sigma = c^*$ as $x \to \infty$ and $\partial\sigma/\partial x = 0$ at $x = 0$, which implies trivially that $\sigma = c^*$.

Equally,

$$\frac{\partial u}{\partial t} = D \frac{\partial^2 u}{\partial x^2}$$

In bulk $u = 2c^*$; at $x = 0$ it follows from substitution that the Nernstian relationships that $c_B$ and $c_C$ can be expressed as functions of $c_A$, and so since $\sigma = c^*$:

$$c_A\left(1 + \exp\left(-\frac{F}{RT}\left(E - E^{\ominus}_{f,A/B}\right)\right)\right.$$

$$\left. + \exp\left(-\frac{2F}{RT}\left(E - \frac{1}{2}\left(E^{\ominus}_{f,A/B} + E^{\ominus}_{f,B/C}\right)\right)\right)\right) = c^*$$

and hence

$$
u = c^* \left( 2 + \exp\left( -\frac{F}{RT}(E - E^{\ominus}_{f,A/B}) \right) \right) \Big/ \left( 1 + \exp\left( -\frac{F}{RT}(E - E^{\ominus}_{f,A/B}) \right) \right.
$$
$$
\left. + \exp\left( -\frac{2F}{RT}\left( E - \frac{1}{2}(E^{\ominus}_{f,A/B} + E^{\ominus}_{f,B/C}) \right) \right) \right)
$$

Therefore $u$ is completely described by equations which do not contain $k_{comp}$ or $k_{disp}$, and since the current is given:

$$
I = -FAD\left( \left.\frac{\partial c_A}{\partial x}\right|_{x=0} - \left.\frac{\partial c_C}{\partial x}\right|_{x=0} \right)
$$
$$
= -FAD\left( 2\left.\frac{\partial c_A}{\partial x}\right|_{x=0} + \left.\frac{\partial c_B}{\partial x}\right|_{x=0} \right)
$$
$$
= -FAD\left( \left.\frac{\partial u}{\partial x}\right|_{x=0} \right)
$$

i.e. as a function of $u$ only, the current too is independent of these parameters. Consequently, under these conditions the voltammetry is 'blind' to the presence or absence of the homogeneous reaction.

(iii) If comproportionation is significant, the species C will not accumulate in solution; rather, it will react with any available A to generate B. The further reduction of B to C at the electrode is rate-limiting, and B will exist in large quantities in the reaction layer. If comproportionation is not significant, B will be rapidly reduced to C and therefore will not exist at a high concentration. The initial reduction of A becomes rate-limiting for the two-electron process. This is made clear in the two concentration profile schematics shown in Fig. 4.9.

Hence, if the presence of B can be determined by an analytical technique, or if it can be determined whether the mass transport of A or B to the electrode is rate-limiting, it is possible to determine whether or not the comproportionation process is active.

(a) C.P. Andrieux *et al.* examined the oxidation of neutral 1,2-ene-diamines in acetonitrile and concluded that comproportionation contributed significantly as they were able to detect the presence of the cation radical by EPR spectroscopy at a potential where its oxidation was fast. If comproportionation was not active, the cation radical would have quickly been consumed by further oxidation and would not have been observable by EPR; as its signal was observed, its regeneration due to the comproportionation reaction was assumed.

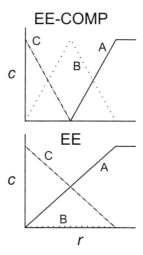

**Fig. 4.9** Comparison of concentration profile schematics for an EE mechanism where the comproportionation reaction is fast (top) or inactive (bottom).

(b) Z. Rongfeng *et al.* studied the reduction of neutral tetracyanoquin-odimethane in acetonitrile where the diffusion coefficients of the different species A, B and C differ significantly enough for the above analysis to be incorrect. The different diffusion coefficients mean that the limiting current due to rate-limiting diffusion of B is distinguishable from that due to rate-limiting diffusion of A.

Using numerical simulation the experimental voltammetry was compared with the predictions for zero and rate-limiting comproportiona-tion, and it was found that the latter gave a superior fit to the experiment.

(c) S.R. Belding *et al.* studied the possible comproportionation of anthraquinone (AQ) in acetonitrile solution. The diffusion coefficients of AQ, its radical anion and its dianion were too similar to permit determi-nation of comproportionation by the above method, within experimental error. The problem was resolved by comparing voltammetry recorded with ample electrolytic support to that recorded under weakly supported conditions.

Under weakly supported conditions, electric fields arising in the solu-tion due to electrolysis are not completely screened and therefore migra-tion contributes to mass transport (see Chapter 10). Since the reduction makes the diffusion layer negative, a species such as neutral AQ is unaf-fected but $AQ^{\cdot-}$ is repelled.

If comproportionation is active, mass transport of $AQ^{\cdot-}$ is rate-limiting, and so the current will be lower due to repulsive migration than if mass transport of AQ is rate-limiting. Again, numerical simulation was used to fit the experimental data, indicating that comproportionation was indeed active. Here, the substantial difference in charges on the different species provided much clearer evidence than the subtle differences in diffusion coefficients.

# 5

---

# Voltammetry at Microelectrodes

## 5.1 Steady-State Concentration Profile in Spherical Space

### Problem

Prove or verify that for a spherical or hemispherical microelectrode of radius $r_e$ under steady-state conditions, the concentration $(c)$ profile of an electroactive species undergoing transport limited electrolysis is

$$c = c^* \left(1 - \frac{r_e}{r}\right)$$

where $c^*$ is the bulk concentration of the species being electrolysed. Hence derive an expression for the transport-limited current to the electrode, for a one-electron oxidation.

### Solution

The steady-state diffusion equation for a spherical or hemispherical microelectrode is

$$\frac{\partial c}{\partial t} = 0 = D\left(\frac{d^2 c}{dr^2} + \frac{2}{r}\frac{dc}{dr}\right) \tag{5.1}$$

The relevant boundary conditions are

$$r = r_e \qquad c = 0$$
$$r \rightarrow \infty \qquad c \rightarrow c^*$$

To *prove* the required result we write $\rho = dc/dr$, so that

$$0 = \frac{d\rho}{dr} + \frac{2}{r}\rho$$

or

$$\frac{d\rho}{\rho} = -\frac{2}{r}dr$$

$$\ln \rho = \ln\left(\frac{1}{r^2}\right) + \ln A$$

where $A$ is a constant. Hence

$$\rho = \frac{dc}{dr} = \frac{A}{r^2}$$

Integrating

$$c = -\frac{A}{r} + B$$

where $B$ is another constant. Applying the boundary condition $c \to c^*$ as $r \to \infty$ requires $B = c^*$. Also, $c = 0$ at $r = r_e$ requires $A = c^* r_e$. Hence

$$c = c^*\left(1 - \frac{r_e}{r}\right)$$

Alternatively, the result may be *verified* by evaluating

$$\frac{dc}{dr} = \frac{c^* r_e}{r^2}$$

and

$$\frac{d^2c}{dr^2} = -\frac{2c^* r_e}{r^3}$$

Substitution into Eq. 5.1 verifies that the suggested concentration profile is indeed a solution to the equation. It also obeys the required boundary conditions.

The transport-limited current for a one-electrode oxidation is

$$I = FAD \left.\frac{dc}{dr}\right|_{r=r_e}$$

where the electrode area $A = 4\pi r_e^2$ (sphere) or $2\pi r_e^2$ (hemisphere). Evaluating the above expression for $dc/dr$ at $r = r_e$ gives

$$\left.\frac{dc}{dr}\right|_{r=r_e} = \frac{c^*}{r_e} \qquad\qquad (5.2)$$

This result shows that the steady-state flux increases with decrease of the electrode radius in accordance with our expectations of convergent diffusion. It follows that

$$I = 4\pi Fc^* Dr_e \qquad \text{(sphere)}$$
$$= 2\pi Fc^* Dr_e \qquad \text{(hemisphere)}$$

Note that the current scales with electrode radius, *not* area.

## 5.2 Current Transients at a Spherical Electrode

### Problem

Following a potential step to a potential where electron transfer is transport-limited, the diffusion-controlled flux $J$, as a function of time, to a spherical electrode of radius $r_e$ is given by

$$J = Dc^* \left[ \frac{1}{\sqrt{D\pi t}} + \frac{1}{r_e} \right]$$

where $c^*$ is the bulk concentration of the electroactive species, and $D$ is its diffusion coefficient.

(i) Sketch how $J$ varies with $t$ in the microelectrode and macroelectrode limits.
(ii) Derive an expression for the time required for the current at a spherical electrode to reach 1% of its steady-state value, and comment on how this varies with $r_e$.

### Solution

(i) For a macroelectrode the term $(1/r_e)$ is tiny and can be neglected, so the current decays to zero according to

$$J = c^* \sqrt{\frac{D}{\pi t}}$$

which is the Cottrell equation for a planar electrode (corresponding to $r_e \rightarrow \infty$). In contrast, for a microelectrode the term $(1/r_e)$ is significant so that the current decays to a finite steady-state value. The curves are illustrated in Fig. 5.1.

(ii) When the current is within 1% of its steady-state value

$$Dc^* \left[ \frac{1}{\sqrt{\pi D t}} + \frac{1}{r_e} \right] = 1.01 \frac{Dc^*}{r_e}$$

$$\therefore \quad \frac{1}{\sqrt{\pi D t}} = 0.01 \frac{1}{r_e}$$

Hence

$$t = \frac{r_e^2 \times 10^4}{\pi D}$$

It follows that the larger the microelectrode, the longer it takes for a steady-state current to be established. For a macroelectrode with $r_e = 1\,\text{mm}$ and $D = 10^{-5}\,\text{cm}^2\,\text{s}^{-1}$ (a typical value), $t \approx 1$ month! Natural convection is likely to influence electrochemical responses after around 20s, so the steady-state regime is never observed for large electrodes.

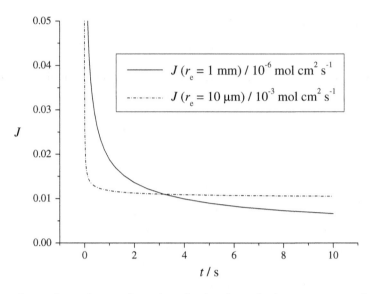

**Fig. 5.1** Comparison of macroelectrode and microelectrode chronoamperometric transients.

## 5.3 Linear vs Convergent Diffusion

### Problem

Show that the ratio of diffusion-limited currents between a spherical electrode of radius $r_e$ and a planar electrode of the same area is given by:

$$\frac{I_{\text{sphere}}}{I_{\text{planar}}} = 1 + \frac{\sqrt{\pi D t}}{r_e}$$

where $D$ is the diffusion coefficient of the electroactive species and $t$ is time. Comment on the physical significance of the result for a spherical electrode.

### Solution

The diffusion-limited current to a planar electrode is given by the Cottrell equation:

$$I_{\text{planar}} = \frac{nFA\sqrt{D}c^*}{\sqrt{\pi t}}$$

where $A$ is the electrode area and $c^*$ is the bulk concentration of the electroactive species. $n$ is the number of electrons transferred per electrolytic event.

The corresponding expression for a spherical electrode is

$$I_{\text{sphere}} = nFADc^* \left( \frac{1}{\sqrt{D \pi t}} + \frac{1}{r_e} \right)$$

Dividing the equations

$$\frac{I_{sphere}}{I_{planar}} = 1 + \frac{\sqrt{\pi D t}}{r_e}$$

as required.

The physical significance of the result is as follows. For short times where $\pi D t \ll r_e^2$, the size of the spherical electrode greatly exceeds that of the diffusion layer, and hence the former experiences effectively planar diffusion (Fig. 5.2).

On the other hand, when $\pi D t \gg r_e^2$, the current at the spherical electrode approaches steady state due to convergent diffusion. The size of the diffusion layer greatly exceeds that of the electrode. Note that the ratio of currents tends to infinity as the time gets longer, since whereas convergent diffusion sustains a steady-state current at the spherical electrode, the current at the planar electrode, which necessarily experiences only planar diffusion, collapses to zero.

## 5.4 Dissolution of Microparticles

### Problem

Small particles of calcite (calcium carbonate, $CaCO_3$) dissolve in acidic solution by means of the following mechanism:

$$H^+_{(aq)} + CaCO_{3(s)} \xrightarrow{k_1} Ca^{2+}_{(aq)} + HCO^-_{3(aq)}$$

$$HCO^-_{3(aq)} + H^+_{(aq)} \rightleftharpoons H_2CO_{3(aq)}$$

$$H_2CO_{3(aq)} \rightarrow H_2O_{(l)} + CO_{2(g)}$$

The following rate law has been measured [R.G. Compton *et al.*, *Freshwater Biology* 22 (1989) 285] for the reaction of protons at the calcite surface:

$$J_{Ca^{2+}}/\mathrm{mol\,cm^{-2}\,s^{-1}} = k_1[H^+]_0 \tag{5.3}$$

where $k_1 = 0.043\ \mathrm{cm\,s^{-1}}$ at 25°C.

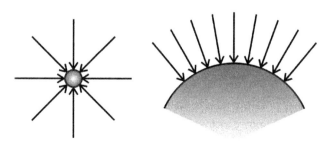

**Fig. 5.2** Schematic of convergent and linear types of diffusion towards spherical electrodes of different sizes.

Explain the form of the rate equation and suggest an explanation for the observation that while large ($>100\,\mu$m) particles of calcite dissolve with a rate controlled by the diffusion of protons to the calcite surface, for much smaller particles the surface controlled reaction quantifed by Eq. 5.3 is found to apply. Assume a value of $7.5 \times 10^{-5}$ cm$^2$ s$^{-1}$ for the diffusion coefficient of H$^+$ in aqueous solution.

## Solution

The rate equation shows that the flux, $J_{Ca^{2+}}$, of dissolving Ca$^{2+}$ ions depends on the concentration of protons, $[H^+]_0$, adjacent to the particle surface (hence the subscript $_0$ on the concentration), in a first-order manner. Given the units of flux are mol cm$^{-2}$ s$^{-1}$ and those of concentration are mol cm$^{-3}$, the units of the first-order rate constant $k_1$ are cm s$^{-1}$.

Whether the dissolution reaction will be controlled by diffusion or by the surface reaction kinetics depends on the relative rates of the two processes. If we assume the CaCO$_3$ particles to be approximately spherical then (using Eq. 5.2) the diffusion-controlled flux of protons to the surface can be calculated from

$$D \left.\frac{\partial [H^+]}{\partial r}\right|_{r=r_e} = \frac{D}{r_e}[H^+]_{\text{bulk}}$$

where $r_e$ is the particle radius. The quantity

$$k_{\text{MT}} = \frac{D}{r_e}$$

is a mass transport coefficient with units of cm s$^{-1}$. If $k_1 \gg k_{\text{MT}}$ then the reaction will be diffusion controlled, whereas if $k_1 \ll k_{\text{MT}}$, it will be surface controlled.

Setting $k_{\text{MT}} = k_1$, we see that

$$k_{\text{MT}} = \frac{D}{r_e} = k_1$$

$$r_e = \frac{D}{k_1} = \frac{7.5 \times 10^{-5}}{0.045}$$

$$\approx 2 \times 10^{-3}\,\text{cm}$$

$$\approx 20\,\mu\text{m}$$

Thus for particles of radius 20 $\mu$m, we would expected 'mixed' kinetics with the surface and mass transport processes operating at similar rates. For CaCO$_3$ particles larger than 100 $\mu$m the processes will be essentially transport controlled, whereas for $r_e \ll 10\,\mu$m the surface reaction will control the overall kinetics.

Calculations of this type are important in gauging the size of $CaCO_3$ particles used for adding to freshwater systems to alleviate the effects of 'acid rain' (so-called 'lake liming').

## 5.5 Steady-State Limiting Current at a Microdisc

### Problem

The incident transport-limited flux at any radius, $r$, on the surface of a microdisc electrode under steady-state diffusion conditions is given by

$$J = \frac{2}{\pi} \frac{c^* D}{\sqrt{r_e^2 - r^2}}$$

where $r_e$ is the disc radius, $c^*$ is the bulk concentration of the electroactive species and $D$ is its diffusion coefficient.

Use this equation to predict the steady-state limiting current at the microdisc electrode for an $n$-electron process. Comment on any possible limitations of the expression for $J$.

### Solution

The current is given by

$$I = nF \int_0^{r_e} 2\pi r \, J \, dr$$

$$= 4nFc^* D \int_0^{r_e} \frac{r \, dr}{\sqrt{r_e^2 - r^2}}$$

Substituting $r = r_e \sin\theta$ and $dr = r_e \cos\theta \, d\theta$

$$I = 4nFc^* D r_e \int_0^{\frac{\pi}{2}} \frac{\sin\theta \cos\theta \, d\theta}{\sqrt{1 - \sin^2\theta}}$$

$$= 4nFc^* D r_e \int_0^{\frac{\pi}{2}} \sin\theta \, d\theta$$

$$= 4nFc^* D r_e [-\cos\theta]_0^{\frac{\pi}{2}}$$

$$= 4nFc^* D r_e$$

The expression for $J$ is derived by assuming a zero concentration of the electroactive species on the disc surface, corresponding to diffusion control. However, the expression predicts the flux to become infinite at the disc edges:

$$J \to \infty \quad \text{as} \quad r \to r_e$$

In practice, this is physically unrealistic since finite electrode kinetics must restrict the flux to a finite value in the vicinity of the disc edge.

## 5.6 Microdisc vs Planar Electrode

### Problem

Consider a microdisc electrode of radius $r_e = 1$ $\mu$m under steady-state conditions for the electrolysis of a species of bulk concentration $c^*$ and diffusion coefficient $D = 10^{-5}$ cm$^2$ s$^{-1}$.

If the same system were electrolysed at a large planar electrode under diffusion-controlled conditions, at what time after the start of electrolysis would the same average flux be observed on the planar electrode as at steady state on the microdisc electrode?

### Solution

The steady-state diffusion-controlled current at the microdisc electrode is given by

$$I = 4nFc^* Dr_e$$

where $n$ is the number of electrons transferred. The corresponding (transient) current at a planar electrode is given by the Cottrell equation:

$$I = \frac{nFA\sqrt{D}c^*}{\sqrt{\pi t}}$$

where $t$ is time and $A$ is the electrode area. Setting $A = \pi r_e^2$ and equating, we find for $D = 10^{-5}$ cm$^2$ s$^{-1}$ and $r_e = 10^{-4}$ cm that $t = 2 \times 10^{-4}$ s. This time is too short to be observed under usual voltammetric conditions at planar electrodes since double layer charging effects are likely to mask the current. It can be appreciated that remarkably high fluxes (current densities) are established at steady state under convergent diffusion conditions, as compared to linear diffusion at planar electrodes.

## 5.7 The Shoup–Szabo Equation

### Problem

The Shoup–Szabo expression gives the current transient at a microdisc electrode of radius $r_e$ resulting from a step from a potential at which no current flows to a potential where the $n$-electron electrolysis of a species is diffusion controlled, the species having bulk concentration $c^*$ and diffusion coefficient $D$. It is:

$$I = 4nFc^* Dr_e f(\tau)$$

where $\tau = 4Dt/r_e$ and

$$f(\tau) = 0.7854 + 0.8862\,\tau^{-\frac{1}{2}} + 0.2146 \exp\!\left(-0.7823\tau^{-\frac{1}{2}}\right)$$

Find and comment on (i) the short and (ii) the long time limits of this expression.

## Solution

(i) As $\tau \to 0$:

$$f(\tau) \to 0.8862\tau^{-\frac{1}{2}}$$

and

$$I \to \frac{nFc^* Dr_e \times 4 \times 0.8862}{(4t/r_e^2)^{\frac{1}{2}}}$$

$$\to \frac{nFc^* D\pi r_e^2}{\sqrt{t}} \times \frac{2 \times 0.8862}{\pi}$$

But $(2 \times 0.8862) = \sqrt{\pi}$, so that

$$I \to \frac{nFA\sqrt{D}c^*}{\sqrt{\pi t}}$$

where $A = \pi r_e^2$ is the disc area. This equation is the Cottrell equation describing linear diffusion to the disc surface at very short times, before the development of any convergent diffusion to the microdisc electrode.

(ii) As $\tau \to \infty$:

$$f(\tau) \to 0.7854 + 0.2146$$

$$\to 1$$

as the exponential term in the expression for $f(\tau)$ becomes unity in the long time limit:

$$\exp\!\left(-0.7823\,\tau^{-\frac{1}{2}}\right) \to \exp(0) = 1$$

Hence

$$I \to 4nFc^* Dr_e$$

which is the expected steady-state current for a microdisc under convergent diffusion.

## 5.8 Steady-State Electrolysis

### Problem

Calculate the transport-limited current at a microdisc electrode of radius $10\,\mu m$ due to the one-electron oxidation of 1 mM ferrocene in acetonitrile/0.1 M tetrabutylammonium perchlorate at 25°C, where the ferrocene has a diffusion coefficient of $2.3 \times 10^{-5}\,cm^2\,s^{-1}$. If the electrode radius were halved, how would the limiting current change, and why? If the cell contained 100 mL of solution, what duration of electrolysis would be required to oxidise 10% of the ferrocene in the cell?

### Solution

The transport-limited current to a microdisc electrode of radius $r_e$ is given for a one-electron oxidation by

$$I = 4Fc^*Dr_e$$

where $D$ is the diffusion coefficient of the electroactive species, $c^*$ its bulk concentration and $F$ is the Faraday constant ($96485\,C\,mol^{-1}$). Hence for the solution in question:

$$I = 4 \times 96485 \times 10^{-6} \times 2.3 \times 10^{-5} \times 10^{-3}$$

where we have used $1\,mM = 10^{-6}\,mol\,cm^{-3}$ and $10\,\mu m = 10^{-3}\,cm$.
    Evaluating:

$$I = 8.9 \times 10^{-9}A$$

$$= 8.9\,nA$$

If the electrode radius is halved then the limiting current is also halved, since the current scales with the electrode radius and *not* with the electrode area. The reason for this lies in the non-uniform current density across the disc electrode surface with a much greater current density at the circumference of the disc than near the disc centre.

    To evaluate the time taken for 10% electrolysis of the cell contents, we first note that 100 mL of a 1 mM solution contains $10^{-4}$ moles of ferrocene. For 10% oxidation we need to calculate the time to electrolyse $10^{-5}$ moles. The required quantity of charge is

$$10^{-5} \times F = 0.96485\,C$$

Hence the required time for 10% electrolysis is

$$t = \frac{0.96485}{8.9 \times 10^{-9}}$$

$$= 1.1 \times 10^8 s$$

This corresponds to 3.5 years! Of course this neglects convection effects, but it is nonetheless true that although diffusional rates and current densities are large at microelectrodes, absolute currents are small and the electrolysis essentially does not perturb the cell contents. If electrosynthesis is to be attempted, a large electrode (and probably large concentrations) are essential.

## 5.9 Effect of Unequal Diffusion Coefficients

### Problem

For an electrochemically reversible wave measured at a microdisc electrode for the process

$$Red - ne^- \rightleftharpoons Ox$$

the half-wave potential, $E_{1/2}$, is related to the formal potential, $E_f^\ominus$, by the expression

$$E_{1/2} = E_f^\ominus + \frac{RT}{nF} \ln \frac{D_{Red}}{D_{Ox}}$$

where $D_{Red}$ and $D_{Ox}$ are the diffusion coefficients of Red and Ox, respectively. For the ferrocene/ferrocenium couple in the room temperature ionic liquid [C4mpyrr][NTf$_2$], $D_{Red} = 2.31 \times 10^{-7}$ cm$^2$ s$^{-1}$ and $D_{Ox} = 1.55 \times 10^{-7}$ cm$^2$ s$^{-1}$.
Calculate $(E_{1/2} - E_f^\ominus)$ and comment on the significance of this result.

### Solution

$$E_{1/2} - E_f^\ominus = +\frac{RT}{F} \ln \frac{2.31 \times 10^{-7}}{1.55 \times 10^{-7}} = 0.010 \, V$$

$$= 8 \, mV$$

The difference of 10 mV is significant since the ferrocene/ferrocenium redox couple is often used as a redox marker in non-aqueous electrochemistry (acting as an internal reference electrode), including that using room temperature ionic liquids as solvents. The correction for the difference in diffusion coefficients is necessary to give a valid reference scale.

## 5.10 Temperature Effects on Steady-State Currents

### Problem

The transport-limited current for an $n$-electron process at a microdisc electrode of radius $r_e$ is given by

$$I = 4nFc^*Dr_e$$

where $D$ is the diffusion coefficient of the electroactive species and $c^*$ is its bulk concentration. $F$ is the Faraday constant.

Does the transport-limited current depend on temperature? If so, why?

## Solution

Although temperature does not appear explicitly in the equation, a change of temperature will indirectly influence three parameters in the equation:

- The concentration of the solution may change slightly as the solution expands (increases its volume) with increasing temperature;
- The radius of the electrode may change slightly with temperature as the metal expands; and
- The diffusion coefficient, $D$, will alter markedly with temperature.

The strong dependence of $D$ on the absolute temperature $T$ (K) derives from the following Arrhenius-type expression:

$$D = D_\infty \exp\left(\frac{-E_a}{RT}\right)$$

where $D_\infty$ is the hypothetical diffusion coefficient at infinite temperature, and $E_a$ is an activation energy for the diffusion process. For ferrocene, $E_a$ has been measured as $6.9\,\text{kJ mol}^{-1}$ in acetonitrile, and $14.4\,\text{kJ mol}^{-1}$ in dimethyl formamide [S.R. Jacob et al., *J. Phys. Chem. B* **103** (1999) 2963]. For N,N,N',N'-tetramethylphenylenediamine (TMPD) in water, $E_a$ was measured as $19.0\,\text{kJ mol}^{-1}$.

Tables 5.1, 5.2 and 5.3 show how the diffusion coefficients for ferrocene in acetonitrile and in DMF (supported by tetra-*n*-butylammonium perchlorate (TBAP)), and for TMPD in water (supported by KCl), will change with temperature, as reported by S.R. Jacob et al., in the above-cited article.

The high sensitivity of diffusion coefficients to temperature is one reason why electrochemical and electroanalytical experiments are carried out under thermostated conditions. Another reason is that electrochemical rate constants also change with temperature, as do formal electrode potentials.

**Table 5.1** Ferrocene in acetonitrile (0.1 M TBAP); error $\pm 1.3 \times 10^{-6}$.

| $T/^\circ$C | 24 | 30 | 40 | 50 | 60 |
|---|---|---|---|---|---|
| $D/10^{-5}\,\text{cm}^2\,\text{s}^{-1}$ | 2.37 | 2.51 | 2.72 | 2.98 | 3.20 |

**Table 5.2** Ferrocene in DMF (0.1 M TBAP); error
$\pm 0.6 \times 10^{-6}$.

| $T/°C$ | 24.8 | 30 | 35 | 40 | 45 |
|---|---|---|---|---|---|
| $D/10^{-5} \, cm^2 \, s^{-1}$ | 1.07 | 1.18 | 1.30 | 1.43 | 1.55 |

**Table 5.3** TMPD in water (0.1 M KCl); error $\pm 0.7 \times 10^{-6}$.

| $T/°C$ | 30 | 40 | 50 | 60 | 70 | 80 |
|---|---|---|---|---|---|---|
| $D/10^{-6} \, cm^2 \, s^{-1}$ | 6.32 | 7.94 | 9.99 | 12.2 | 14.9 | 18.6 |

## 5.11 ECE Mechanism at a Microdisc Electrode

## Problem

For an ECE mechanism

$$A + e^- \rightleftharpoons B$$

$$B \xrightarrow{k} C$$

$$C + e^- \rightleftharpoons products$$

which is studied at a microdisc electrode of radius $r_e$, the effective number of electrons transferred, $n_{eff}$, is very *approximately* given by the following expression [M. Fleischmann *et al.*, *J. Electroanal. Chem.* **177** (1984) 115]:

$$\frac{1}{n_{eff} - 1} = \left[ \frac{4}{\pi} \left( \frac{D}{k} \right)^{\frac{1}{2}} \cdot \frac{1}{r_e} \right] + 1$$

where $D$ is the diffusion coefficient of species A and $k$ is the first-order rate constant for the conversion of B to C.

(i) How and why does $n_{eff}$ vary as $r_e$ changes from a small to a large value?
(ii) For the reduction of *m*-iodonitrobenzene in acetonitrile solution, the following data were obtained for the transport-limited current as a function of microdisc electrode radius [R.G. Compton *et al.*, *Electroanalysis* **8** (1996) 214]:

| $r_e / \mu m$ | 5.6 | 13.6 | 30.0 | 60.0 |
|---|---|---|---|---|
| $I/nA$ | 2.9 | 8.9 | 19.0 | 41.0 |

Assuming a value of $D = 2.1 \times 10^{-5} \, cm^2 \, s^{-1}$, estimate a value for $k$ and suggest a mechanism for the electrode reaction.

## Solution

(i) As $r \to 0$, $\frac{1}{(n_{eff}-1)} \to \infty$ and so $n_{eff} \to 1$.

In the other limit, $r \to \infty$, $\frac{1}{(n_{eff}-1)} \to 1$ and so $n_{eff} \to 2$.

The variation arises since for a very small electrode the convergent/divergent diffusion to and from the electrode is extremely effective, and so species B is removed from the electrode *before* there is time for the species to undergo reaction to form C and then gain a second electron. Hence in the $r_e \to 0$ limit, $n_{eff} \approx 1$.

In contrast, for a very large electrode the diffusional transport is much less effective and so B can remain near the electrode long enough to undergo decomposition into C and hence further reduction; and so in the limit of $r_e \to \infty$, $n_{eff} \to 2$.

(ii) The data can be analysed by trial and error using a spreadsheet. First, the equation above can be used to find $n_{eff}$ for various values of the ratio $(D/k)$ for the different electrode sizes. Second, the absolute current can be calculated using

$$I = 4 n_{eff} F c^* D r_e \tag{5.4}$$

and compared with experiment. Using the value for $D$ given, the rate constant $k = 0.30 \text{ s}^{-1}$ gives the following currents:

| $r_e/\mu m$ | 5.6 | 13.6 | 30.0 | 60.0 |
|---|---|---|---|---|
| $I/nA$ | 3.1 | 8.1 | 19.7 | 41.1 |

which are in good agreement with experiment.

Alternatively, a graphical approach can be taken in which the experimental current values are converted into $n_{eff}$ using Eq. 5.4 and then $(n_{eff} - 1)^{-1}$ is plotted against $r_e^{-1}$. The straight line graph intersects the $y$-axis close to unity and has a slope of $(4/\pi)(D/k)^{\frac{1}{2}}$, allowing the determination of $k$. A likely mechanism is

$$I-C_6H_4NO_2 + e^- \rightleftharpoons [I-C_6H_4NO_2]^{\cdot-}$$

$$[I-C_6H_4NO_2]^{\cdot-} \rightarrow {}^{\cdot}C_6H_4NO_2 + I^-$$

$${}^{\cdot}C_6H_4NO_2 + HS \rightarrow C_6H_5NO_2 + S^{\cdot}$$

$$C_6H_5NO_2 + e^- \rightleftharpoons [C_6H_5NO_2]^{\cdot-}$$

where HS denotes the solvent and/or supporting electrolyte.

Note that digital simulation of the full voltammetric waves has superseded the use of approximate equations in electrochemical research.

## 5.12 EC′ Mechanism at a Microdisc Electrode

### Problem

The simple catalytic (EC′) mechanism follows the scheme below:

$$\text{Red} - ne^- \rightleftharpoons \text{Ox}$$

$$\text{Ox} + \text{Z} \xrightarrow{k} \text{Red} + \text{products}$$

For the case of a microdisc equation the following very *approximate* equation has been derived for the effective number of electrons transferred, $n_{eff}$, as a function of the electrode radius, $r_e$, under steady-state conditions. Ox and Red are assumed to have the same diffusion coefficient $D$ and the homogeneous kinetic step is assumed to show pseudo-first-order kinetics with a rate constant $k' / s^{-1} = k[Z]$.

$$n_{eff} = n\left(1 + \frac{\pi}{4} r_e \left(\frac{k'}{D}\right)^{\frac{1}{2}}\right)$$

The method was used to study the oxidation of ferrocyanide (0.41 mM) in the presence of amidopyrine (2 mM) in aqueous KOH [M. Fleischmann *et al.*, *J. Electroanal. Chem.* **177** (1984) 97]. The following data were determined:

| $n_{eff}$ | 1.04 | 1.22 | 1.42 |
|---|---|---|---|
| $r_e/\mu m$ | 0.25 | 2.5 | 5.0 |

Using a value of $D = 6 \times 10^{-6}$ cm$^2$ s$^{-1}$, estimate a value for $k$ and explain physically why $n_{eff}$ increases as the electrode gets larger.

### Solution

A graph of $n_{eff}$ vs. $r_e$ is a straight line to within experimental error. The intercept corresponds to $n_{eff} = 1$, consistent with the simple E reaction

$$[\text{Fe(CN)}]_6^{4-} - e^- \rightleftharpoons [\text{Fe(CN)}]_6^{3-}$$

The gradient of the graph is

$$\frac{\pi}{4}\left(\frac{k'}{D}\right)^{\frac{1}{2}} = 0.08 \,\mu m^{-1}$$

$$= 800 \text{ cm}^{-1}$$

so that

$$k = \frac{k'}{[\text{amidopyrine}]} = \frac{\left(800 \times \frac{4}{\pi}\right)^2 \times 6 \times 10^{-6}}{2 \times 10^{-3}}$$

$$\approx 3 \times 10^3 \text{ M}^{-1}\text{s}^{-1}$$

The effective number of electrons transferred increases with the electrode size since the efficiency of transport via convergent/divergent diffusion to or from the electrode decreases. Hence, the species Ox spends more time close to the electrode than is the case when the electrode is very small. Accordingly, there is more time available for the reaction with Z to produce Red that will be able to react further at the electrode, and hence a greater catalytic activity.

Note that the above theoretical treatment is highly approximate; the modern approach to data analysis would be via digital simulation allowing for depletion of Z and for variable diffusion coefficients, as well as for providing a rigorous (rather than approximate) solution of the coupled diffusion–kinetic equations involved.

## 5.13  Size Effects on Half-Wave Potentials

### Problem

For an electrochemically irreversible reduction at a microdisc or microsphere electrode of radius $r_e$, the variation of the voltammetric half-wave potential, $E_{1/2}$, under steady-state conditions can be shown to vary as

$$\frac{\partial E_{1/2}}{\partial \ln r_e} = \frac{RT}{\alpha F}$$

where $\alpha$ is the transfer coefficient [F.W. Campbell *et al.*, *J. Phys. Chem. C* 113 (2009) 9053]. The following data were obtained for the reduction of hydrogen peroxide at isolated silver nanoparticles supported on an inert but conducting substrate electrode:

| $r_{NP}$/nm | 15 | 30 | 50 |
|---|---|---|---|
| $E_{pf}$/V | −1.575 | −1.503 | −1.453 |

Explain physically why the peak shifts with the particle radius and estimate a value of $\alpha$. If the substrate were covered with a monolayer of the silver nanoparticles, explain qualitatively what would be seen.

### Solution

The peak shifts with the particle radius such that a greater overpotential is required for transport-limited reduction as the particle shrinks in size. This is because the convergent/divergent regime of mass transport to and from the particle becomes more effective and dominates as the particle shrinks. Accordingly, a greater potential has to be applied to attain the peak current, as faster electrode kinetics are required to compete with mass transport. Hence, $E_{pf}$ shifts to more negative values as $r_{NP}$ decreases.

To find a value of $\alpha$, plot a graph of $E_{pf}$ vs $\log_{10} r_{NP}$. A straight line is seen with a slope of

$$\frac{\partial E_{pf}}{\partial \log_{10} r_{NP}} = \frac{2.3RT}{\alpha F} = 233 \, \text{mV}$$

from which it may be inferred that $\alpha \approx 0.25$. Interestingly, this value is close to that measured at silver macroelectrodes for the reduction of hydrogen peroxide.

If the substrate electrode were fully covered with silver nanoparticles then the entire surface would behave as a macroelectrode with essentially planar diffusion. As a result of the much less effective (planar, not convergent) diffusion, the half-wave potential and peak potential would shift to less negative values.

## 5.14 Extracting Parameters from Microdisc Chronoamperometry

### Problem

A 10 mM aqueous solution of hydrazine, $N_2H_4$, was studied in pH 4.8 acetate buffer using a platinum microdisc electrode of radius 13.8 $\mu$m. Chronoamperometric transients were recorded in which the potential was stepped from a value corresponding to no current flow to one corresponding to the transport-limited current for the oxidation of $N_2H_4$. The following short-time data were recorded:

| $t$/ms | 5.5 | 10.5 | 15.5 | 20.5 |
|---|---|---|---|---|
| $I/10^{-7}$ A | 6.84 | 5.2 | 4.53 | 4.15 |

| $t$/ms | 25.5 | 30.5 | 35.5 | 40.5 |
|---|---|---|---|---|
| $I/10^{-7}$ A | 3.91 | 3.72 | 3.58 | 3.44 |

At long times a steady-state current of $1.96 \times 10^{-7}$ A was measured. Calculate the number of electrons transferred in the oxidation of hydrazine.

### Solution

Since the concentration of the hydrazine is known to be 10 mM, the current-time data and the steady-state current allow us to find two unknowns, namely $n$, the number of electrons transferred in the oxidation, and $D$, the diffusion coefficient of $N_2H_4$.

For the steady-state current

$$I_{\lim} = 4nFc^*Dr_e$$

from which it follows that

$$nD = \frac{I_{\lim}}{4Fc^*r_e}$$

$$= \frac{1.96 \times 10^{-7}}{4 \times 96485 \times 10^{-5} \times 1.38 \times 10^{-3}}$$

$$= 3.68 \times 10^{-5}\,\text{cm}^2\,\text{s}^{-1}$$

where $c^*$ and $r_e$ have been converted to cm units.

The short-time data presented are in the range 5–40 ms, which are long enough times, in aqueous solution, for double layer charging effects to have decayed to a negligible value, but are short enough that linear rather than convergent diffusion to the electrode prevails. Thus if a plot is made of current against $t^{-\frac{1}{2}}$, a Cottrellian slope can be expected.

Note this plot is not expected to go through the origin, because of the decay to a constant value at a microelectrode, although the slope will be Cottrellian at short enough times. The limiting current value is not small enough compared to the current values reported in the table of data for the intercept of a Cottrell plot to be near zero.

The Cottrellian slope is:

$$\text{gradient} = nFAc^*D^{\frac{1}{2}}\pi^{-\frac{1}{2}}$$

where $A$ is the electrode area; for a disc, $A = \pi r_e^2$.

A graph of $I$ vs $t^{-\frac{1}{2}}$ shows a straight line of gradient $3.95 \times 10^{-8}\,\text{A}\,\text{s}^{\frac{1}{2}}$. Therefore

$$nD^{\frac{1}{2}} = \frac{\text{gradient}}{F\pi^{\frac{1}{2}}r_e^2c^*}$$

$$= \frac{3.99 \times 10^{-8}}{96485 \times \pi^{\frac{1}{2}} \times (1.38 \times 10^{-3})^2 \times 10^{-5}}$$

$$= 0.0121\,\text{cm}\,\text{s}^{-\frac{1}{2}}$$

Clearly

$$D = \left(\frac{nD}{nD^{\frac{1}{2}}}\right)^2 = \left(\frac{3.68 \times 10^{-5}}{1.21 \times 10^{-2}}\right)^2 = 9.21 \times 10^{-6}\,\text{cm}^2\,\text{s}^{-1}$$

and hence $n = (36.8/9.21) = 4.00\,(= 4$ to within experimental error).

It follows that the oxidation of hydrazine is a four-electron process.

$$N_2H_{4(aq)} - 4e^- \rightarrow N_{2(g)} + 4H^+_{(aq)}$$

## 5.15 Extracting Parameters from Microdisc Chronoamperometry

### Problem

The electroreduction of an approximately 2 mM solution of nitrobenzene in acetonitrile/0.1 M tetrabutylammonium perchlorate was studied using a platinum microdisc electrode of radius, $r_e$, 24 $\mu$m. A steady-state limiting current of $-4.05 \times 10^{-8}$ A was observed at long time, while at short times the following current ($I$)-time ($t$) data were collected.

| $t/10^{-2}$ s | 1 | 2 | 3 | 4 | 5 |
|---|---|---|---|---|---|
| $-I/10^{-8}$ A | 9.70 | 6.86 | 5.59 | 4.86 | 4.35 |

| $t/10^{-2}$ s | 6 | 7 | 8 | 9 | 10 |
|---|---|---|---|---|---|
| $-I/10^{-8}$ A | 3.94 | 3.67 | 3.43 | 3.22 | 3.06 |

Assuming that the reduction is a one-electron process forming the nitrobenzene radical anion, calculate the diffusion coefficient $D$ of nitrobenzene and obtain an accurate value for the concentration studied.

### Solution

The steady-state current for a reduction is given by:

$$I_{lim} = -4nFc^*Dr_e$$

and so if $n = 1$:

$$c^*D = \frac{-I_{lim}}{4Fr_e}$$

$$= \frac{4.05 \times 10^{-8}}{4 \times 96485 \times 24 \times 10^{-4}}$$

$$= 4.37 \times 10^{-11} \, \text{mol cm}^{-1} \text{s}^{-1}$$

where again $D$, $r_e$ and $c^*$ are in cm units.

A plot of $-I$ vs $t^{-\frac{1}{2}}$ for the short-time data has a slope of $9.71 \times 10^{-9}$ A s$^{\frac{1}{2}}$. This should be equal to the Cottrellian value:

$$\text{gradient} = nFAc^*D^{\frac{1}{2}}\pi^{-\frac{1}{2}}$$

where $n = 1$ and $A = \pi r_e^2$, and so

$$c^* D^{\frac{1}{2}} = \frac{\text{gradient}}{F\pi^{\frac{1}{2}} r_e^2} = \frac{9.71 \times 10^{-9}}{96485 \times \pi^{\frac{1}{2}} \times (24 \times 10^{-4})^2}$$

$$= 9.86 \times 10^{-9} \text{ mol cm}^{-2} \text{ s}^{-\frac{1}{2}}$$

As before we can combine our results to determine

$$D = \left( \frac{c^* D}{c^* D^{\frac{1}{2}}} \right)^2 = \left( \frac{4.37 \times 10^{-11}}{9.86 \times 10^{-9}} \right)^2 = 1.97 \times 10^{-5} \text{ cm}^2 \text{ s}^{-1}$$

and hence $c^* = 2.22 \times 10^{-6} \text{ mol cm}^{-3} = 2.22 \text{ mM}$.

# 6

## Voltammetry at Heterogeneous Surfaces

### 6.1 Graphitic Electrodes

#### Problem

Carbon comes in a variety of allotropes which exhibit a wide range of different properties; graphite is one of the most common of these forms. A recent review by R.E. McCreery [*Chem. Rev.* **108** (2008) 2646] provides a good overview on a number of carbon-based materials and their use in electrochemistry.

(i) What is the physical structure of graphite? What is its electronic structure?
(ii) Why are the electron transfer kinetics for a redox species generally slower on graphitic electrodes as compared to metallic electrodes? Further, why is a difference in electrochemical activity observed for different planes of the graphite?
(iii) Based on your knowledge of the graphite structure, when fabricating a graphite electrode how may we control its reactivity?

#### Solution

(i) Experimentally it is common to use graphite which has been synthetically produced. This graphite is known as *highly ordered pyrolytic graphite* (HOPG). It is used as it has a high purity and its structure has a high degree of three-dimensional ordering. A schematic representation of the structure of HOPG is shown in Fig. 6.1.

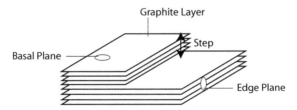

**Fig. 6.1** Schematic representation of highly ordered pyrolytic graphite. Reproduced from C.E. Banks *et al.*, *Chem. Commun.* (2005) 829, with permission from the Royal Society of Chemistry.

Structurally, graphite is comprised of layers: these layers are two-dimensional sheets of hexagonally arranged carbon atoms where the carbon atoms may be described as being $sp^2$ hybridised. These layers are stacked on top of each other with an interlayer spacing of 3.35 Å. Due to the carbon being $sp^2$ hybridised, each carbon atom has a half filled p-orbital which lies perpendicular to the layers. It is the overlap of these p-orbitals which leads to the electrons being delocalised in the interlayer regions. These delocalised electrons are 'free' to move within each layer but there is no direct contact between the layers, and, consequently, the electrical properties of graphite are anisotropic.

Electronically, graphite may be described as a semi-metal. This is as a result of its substantially lower density of states, as compared to that of a metal. Graphite does not exhibit a band gap as there is overlap between the conduction and valence bands, and hence there is a non-zero density of states at the Fermi level.

(ii) The rate of electron transfer is described by the Marcus–Hush formalisation for electron transfer. Through utilisation of Fermi's 'golden rule', this formalisation has been extended to being fully quantum mechanical in nature. Such extended forms highlight how the rate of electron transfer is in part determined by the overlap integral between the molecular orbitals of the redox species and the electronic energy levels present within the electrode. The density of states for graphite is substantially lower than that of a metallic material, and so the overlap integral for a graphitic electrode is lower and hence the rate of electron transfer is commonly observed to be slower. Recent work by N.S. Lewis *et al.* [*J. Phys. Chem. B* **110** (2006) 19433] has focused on the difference in rates of electron transfer at metallic and graphitic electrodes from a fundamental standpoint.

As mentioned above, the electronic structure of graphite is anisotropic, and this anisotropy is reflected in the differing electrical resistance of the material perpendicular and parallel to the carbon layers. Further, as a result of this anisotropy the 'edge' and 'basal' plane sites (as indicated in Fig. 6.1) exhibit significantly different electrochemical properties. The rate of electron transfer is generally accepted to be orders of magnitude greater at the edge plane sites; for more information, see the work of C.E. Banks and R.G. Compton [*Analyst* **131** (2006) 15].

(iii) As has been described in parts (i) and (ii) of this problem, carbon is anisotropic, with the 'edge' and 'basal' plane sites exhibiting significantly different electrochemical responses. Consequently, when fabricating an electrode the experimentalist may choose the alignment of the graphitic material, and as such may produce an electrode which has either a higher surface coverage of either 'basal' or 'edge' plane sites. As a point of caution, although the material may be aligned differently the resulting electrodes will not be purely 'edge' or 'basal' in character, due to inherent defects present. This is explored further in Problem 6.3.

## 6.2 Carbon Nanotubes and Their Reactivity

### Problem

Since their rediscovery in 1991 by S. Iijima [*Nature* **354** (1991) 56], carbon nanotubes have been the focus of a huge amount of research.

(i) A number of varieties of carbon nanotubes are available; briefly summarise their structures.

(ii) The electrochemical response of carbon nanotubes are regularly compared to macroscopic graphite. Explain the concepts behind these analogies making reference to both the edge and basal plane sites present on graphite.

(iii) Carbon nanotubes have been experimentally observed to be 'electrocatalytic' for the oxidation and reduction of a vast number of species. Explain how metallic impurities present within the carbon nanotubes may influence the observed voltammetry.

### Solution

(i) Carbon nanotubes may be divided into two main groups: single-walled carbon nanotubes (SWCNT) and multi-walled carbon nanotubes (MWCNT). MWCNTs come in a number of forms, including hollow, bamboo and herringbone. The structures of these three are shown schematically in Fig. 6.2. SWCNTs may be understood as being a single sheet of graphene which has been rolled into a tube.

(ii) A comparison is often made between macroscopic graphite and multi-walled carbon nanotubes in order to help to qualitatively explain the observed

| Hollow | Herringbone | Bamboo |

**Fig. 6.2** Schematic representation of the cross-section of the three main forms of multi-walled carbon nanotubes.

voltammetry at carbon nanotubes, where the sidewalls of the carbon nanotubes are seen as being analogous to the basal plane sites on graphite, and the tube ends are viewed as being comparable to the edge plane sites. This simplification leads to a qualitative explanation for the observed voltammetry, but it should be noted that there are limitations to this analogy due to the fact that the energies associated with the carbon nanotubes will be significantly altered from those for macroscopic graphite, and further the density of states for carbon nanotubes will be significantly different.

(iii) Carbon nanotubes may be synthesised *via* a number of different methods, but one of the most common routes for their production is through the use of *chemical vapour deposition* (CVD). CVD uses a nanoparticulate metal such as iron (or a combination of metals including cobalt and nickel) to catalyse their growth. These metallic nanoparticles, or nanoparticles of the corresponding metal oxides, may then become occluded within the carbon nanotube. Consequently, the observed 'electrocatalytic' effect of the carbon nanotubes may in a number of cases be simply due to the occluded metal and not the nanotube itself. For further reading on this subject see C.E. Banks *et al.* [*Angew. Chem.* **45** (2006) 2533] or M. Pumera *et al.* [*TrAC 9* (2005) 177].

## 6.3 Highly Ordered Pyrolytic Graphite and the Influence of Defects

### Problem

Through the use of *highly ordered pyrolytic graphite* (HOPG) it is possible to produce an electrode which is predominantly 'basal' in character, but even with such a surface, edge plane defects will be present in the form of 'steps' (as indicated in Fig. 6.1). Careful preparation can lead to a surface where these edge plane defects are up to 1–10 $\mu$M apart.

(i) From your knowledge of the electrochemistry of carbon surfaces, explain how these edge plane steps can dominate the observed voltammetry.

(ii) Suggest why, when modelling such systems, the use of a one-dimensional diffusion model is inappropriate.

# Solution

(i) As discussed in Problem 6.1, the electrochemical activity of the basal and edge plane sites is significantly different: the edge plane sites generally exhibit far higher rates of electron transfer. As a consequence of these higher rates of electron transfer, the observed voltammetry is dominated by the surface coverage of the edge plane defects see C.M. Neuman *et al.* [*Chem. Eur. J.* **17** (2011) 7320].

(ii) A one-dimensional model requires that the rate of electron transfer is uniform across the whole of the electrode, but this is clearly not the case. At first sight, we might assume that the average rate of electron transfer ($k_{obs}$) could be described by the following equation,

$$k_{obs} = \Theta_{edge} k_{edge} + \Theta_{basal} k_{basal}$$

where $\Theta_i$ and $k_i$ are, respectively, the fractional surface coverage and rate of electron transfer for the site $i$. However, such an approximation would be sufficient only if strictly linear diffusion is occurring to the electrode surface, which is not the case. Rather, as a result of the relatively large spacing between the edge plane defects, non-linear diffusion will play a significant role in transport to the more active edge plane defects (steps) on the surface.

Consequently, the use of a two- or three-dimensional simulation is required to successfully model this system. Further reading on this subject can be found in the work of C.E. Banks *et al.* [*Chem. Comm.* **7** (2005) 829].

## 6.4 Advantages of Arrays

### Problem

Why might it be advantageous to use a microelectrode array in place of either a single microelectrode or a macroelectrode of the same material?

### Solution

Compared to a single microelectrode, a microelectrode array has the obvious advantage that the current scales with the number of electrodes, provided the electrodes are diffusionally independent. For analytical purposes, the signal strength of a microelectrode array is therefore much larger and so lower concentrations can be detected accurately. So long as the array is well designed, in terms of the size and spacing of electrodes (as discussed in Problem 6.5), it will retain the advantageous convergent diffusion of a microelectrode, and so will exhibit a steady-state current and fast rates of diffusion.

As discussed in Chapter 5, these properties facilitate the determination of fundamental parameters such as diffusion coefficients or rate constants, as compared to a macroelectrode. What is more, the electroactive area of a microelectrode array is much less than that of a macroelectrode, and so the capacitive current is markedly reduced. The signal is dominated by Faradaic processes (electrochemical reactions), thereby facilitating data analysis.

The reduced electroactive area of a microelectrode array is also economical. Frequently, electrode materials are noble metals such as platinum or gold, since these are stable over a wide potential window and frequently exhibit fast electrode kinetics or catalysis towards common analytes. These metals are necessarily very expensive, but by using a microelectrode array in place of a macroelectrode, the required quantity of the electrode material can be greatly reduced, in addition to the electroanalytical advantages noted above.

## 6.5  Diffusional 'Cases'

### Problem

The work of T.J. Davies *et al.* [*J. Solid State Electrochem.* **9** (2005) 797] identified four diffusional cases which arise at microelectrode arrays. As time passes and diffusion layers grow, the behaviour changes from one to the next. The cases are:

Case 1: Predominantly planar diffusion to isolated electrodes.
Case 2: Predominantly convergent (radial) diffusion to isolated electrodes.
Case 3: Limited overlap between diffusion layers to neighbouring electrodes.
Case 4: Near complete overlap between diffusion layers to neighbouring electrodes; overall planar diffusion layer to the array.

> (i) For each case, explain what a typical voltammogram will look like and what equations, if any, may be used to analyse chronoamperometry and cyclic voltammetry at this timescale.
>
> (ii) Why, in practice, is ideal Case 4 behaviour sometimes not observed even at a densely packed array?

### Solution

(i) Case 1: If the electrode is isolated and the timescale is sufficiently short that diffusion can be considered planar, then the analysis in Chapter 4 applies. The Cottrell equation can be used for chronoamperometry ($I \propto t^{-\frac{1}{2}}$) and the Randles–Ševčík equation can be used for cyclic voltammetry ($I_{pf} \propto v^{\frac{1}{2}}$), in which the area, $A$, in these equations is the total electroactive area of the array. We expect the voltammetry to be 'peak shaped'.

Case 2: If the electrode is isolated but the timescale is long enough that diffusion is convergent, the analysis in Chapter 5 applies. Therefore, a chrono-amperometric transient tends to a steady-state current ($I_{ss} = 4nFc^*Dr_e$

for a microdisc), and the voltammogram will be sigmoidal with negligible hysteresis, tending to an equivalent steady-state current at high overpotential.

Case 3: If diffusion layers overlap, neighbouring electrodes will shield each other's diffusion zones. Therefore, the current at each electrode will be less than predicted by the steady-state equation for Case 2. Additionally, the voltammetry is slightly peak-shaped, since depletion at the boundaries between diffusional zones associated with each electrode will limit diffusion-controlled current. No simple analytical expressions exist for this regime, and therefore numerical simulation is essential.

Case 4: Once overlap is total between diffusion layers, the behaviour of an individual electrode no longer impacts upon the voltammetry. Rather, diffusion occurs to the array as a whole. Since the array is typically macroscale, we can use the Cottrell equation and Randles–Ševčík equations as in Case 1. However, the area is now the total array area $A_{array}$ rather than just the area of the electroactive components.

Additionally, the apparent heterogeneous rate constant is altered in Case 4, since the size and distribution of the electrodes where the reaction is taking place does not correlate directly to the overall planar electrode assumed with a one-dimensional theory. According to the theory due to C. Amatore *et al.* [*J. Electroanal. Chem.* **147** (1983) 39], the apparent $k^0$ on the basis of assuming a homogeneous surface is given by $k_{eff}^0 = k^0(1 - \Theta)$ where $\Theta$ is the surface coverage.

(ii) Case 4 behaviour requires that the electrodes in a dense array are quite regularly distributed and that they are electroactive. If, as may often occur in practice, the electrodes have been clustered in a random array due to a poor lithographic manufacturing technique, or 'dead' (unconnected) electrodes occur in blocks rather than in a random distribution, there may be significant areas in which no electroactive sites exist, even if the overall coverage is relatively high.

In this case, the gaps between electroactive sites will not be filled by expanding diffusion layers as quickly as in areas where all electrodes are active, and so the current is proportionally reduced from that predicted by Case 4 theory. See also the discussion on partially blocked electrodes (Problem 6.8).

## 6.6 Geometry of a Regular Array

### Problem

Suppose that a microdisc array consists of a hexagonal array of microdiscs separated by insulating material, each disc having radius $r_e = 1 \, \mu m$, and a line-of-centres separation between discs of $10 \, \mu m$.

(i) Calculate the area of one unit cell.
(ii) What is the proportional coverage of electrode material?
(iii) Using the diffusion domain approximation, at what radius in a cylindrical simulation space should a symmetry boundary be applied?
(iv) Assuming a diffusion coefficient of $D = 10^{-5}$ cm$^2$ s$^{-1}$, on what timescales might each of the behaviours of diffusional Cases 1 to 4 be expected?

## Solution

(i) Each unit cell is a hexagon with a centre-to-edge distance of $h = 5\,\mu$m (or, equivalently, the rhombic unit cell in Fig. 2.2). The area of the unit cell is six times the area of an equilateral triangle with this height and base $b$:

$$A_{cell} = 6 \times \frac{bh}{2} = 6 \times \frac{1}{2} \times h \times \left(2h \tan\left(\frac{\pi}{6}\right)\right)$$
$$= 2\sqrt{3} \times h^2$$
$$= 86.6\,\mu\text{m}^2 = 8.66 \times 10^{-11}\,\text{m}^2$$

(ii) The coverage, $\Theta$, is the ratio of electroactive surface area to total surface area, or, equivalently, the area of one disc electrode to the area of its unit cell. Since the area of the disc electrode is simply $A_{el} = \pi r_e^2$, the coverage is:

$$\Theta = \frac{\pi \times 1^2}{86.6} = 0.036$$

i.e. about 3.6% of the total surface area is electroactive.

(iii) Taking into account distance perpendicular from the electrode array (the $z$-axis), the unit cell is a hexagonal prism which does not have a symmetry that can be simulated by just a two-dimensional plane. Therefore, the diffusion domain approximation is used to approximate this unit cell to a cylinder with an equivalent basal area. If the diffusion domain has a radius $r_{DD}$:

$$\pi r_{DD}^2 = A_{cell}$$

and so

$$r_{DD} = \sqrt{\frac{A_{cell}}{\pi}}$$
$$= 5.25\,\mu\text{m}$$

If we apply a zero-flux boundary at $r = r_{DD}$, it is approximately equivalent to simulating an infinite hexagonal array described as above, but the simulation involves a two-dimensional space rather than a three-dimensional space, and therefore is much faster. Extensive studies have shown this approximation to be accurate for the purpose of analysing many experiments.

(iv) The transition from Case 1 to Case 2 occurs when the radial term in diffusion becomes dominant. According to the theories set out in Chapter 5, this arises when

$$t \gg \frac{r_e^2}{D} \approx 1\,\text{ms}$$

Therefore, Case 1 behaviour (the Cottrell/Randles–Ševčík regime) is only significant for this example at timescales less than a millisecond, as would be expected for a microdisc electrode.

The transition from Case 2 to Case 3 occurs with diffusion layer overlap, which will occur after diffusion layers from each electrode have extended about 5.25 $\mu$m and therefore encountered each other and begun to shield each other's diffusion zones. Since the mean diffusion layer extent in any direction is $x_{\text{diff}} \approx \sqrt{2Dt}$, to diffuse a characteristic distance of 5.25 $\mu$m requires:

$$t \approx \frac{x_{\text{diff}}^2}{2D} \approx 14\,\mu\text{s}$$

Therefore, a true Case 2 behaviour (steady-state voltammetry to an isolated microdisc) is not likely to be observed, since diffusion layers will interact for this array after only tens of microseconds.

When diffusion layers overlap by a large amount, an overall planar response will be expected, but with a characteristic area equivalent to the total array surface area rather than just the electroactive surface area. Hence, the Case 4 current will be $(1/\Theta)$ times larger than the Case 1 current. This will occur when $x_{\text{diff}} \gg d$ where $d$ is the separation of the individual microdiscs. There-fore, Case 4 behaviour arises at $t \gg 0.1$ s. This will therefore be the domi-nant behaviour for cyclic voltammetry at normal scan rates at this particular array. With chronoamperometry, short timescales are accessible and so Case 3 behaviour may also be observed.

## 6.7 Analysis of Diffusion to Electrode Arrays

## Problem

(i) I. Streeter *et al.* studied BPPG electrodes modified with palladium-covered car-bon microspheres [*J. Phys. Chem. C* **111** (2007) 17008]. The results showed that for a fractional coverage of $\Theta = 0.445$, voltammetry at the array obeyed the Randles–Ševčík equation exactly, but for $\Theta = 0.115$ and below, the Randles–Ševčík equation was obeyed only for scan rates less than 50 mV s$^{-1}$. Explain these results.

(ii) In their study of the activity of highly ordered pyrolytic graphite electrodes (see Problem 6.3), C.E. Banks *et al.* showed that for voltammetry at a basal plane HOPG electrode, a 'best fit', using the one-dimensional diffusion program DigiSim, underestimated the diffusional tail of the forward peak and overestimated the magnitude of the reverse peak. It was proposed that this showed that only nanoscale 'edge plane' sites separating the basal planes were electroactive – how does this explain the observed voltammetry?

## Solution

(i) When a BPPG electrode is modified with a high coverage of electroactive palladium-covered carbon microspheres, which may be assumed to be randomly distributed, the spacing between neighbouring electroactive sites is short with respect to the distance diffused by the reactant in a typical cyclic voltammetry experiment. Therefore, individual diffusion zones do not impact on the voltammetry, and the diffusion layer is approximately planar due to overlap of diffusion zones from neighbouring microspheres. This is Case 4 behaviour – the Randles–Ševčík equation is obeyed.

For lower coverages at fast scan rates, the overlap between neighbouring diffusion zones is not total because these are more distantly spaced than for high coverage. Therefore, the diffusion layer is not ideally planar and a lower current is observed than expected from the Randles–Ševčík equation since some material between the electrodes will not react. At slower scan rates, however, the diffusion zone is larger since the experimental timescale is longer, and so fuller overlap is attained. Therefore we see a scan rate dependent transition from Case 3 to Case 4 behaviour.

(ii) The enhanced diffusional tail in the forward peak arises because the active edge plane sites are much smaller than the electrode itself, as would be simulated in DigiSim. Therefore, DigiSim does not consider the elevated rate of mass transfer to the very small edge plane sites, in which convergent diffusion is much faster than the planar diffusion towards the electrode as a whole. As a consequence, the kinetics of the electron transfer process may become rate-limiting at the edge plane sites themselves. This leads to overall irreversibility and an altered waveshape from the best fit using DigiSim.

The loss of current in the reverse peak reflects Case 4 behaviour: some material is likely to be lost to the region between edge plane sites during the scan, and because electroactive material is not present across the entire electrode, it is possible that some of this material is not converted back to starting material during the scan. Additionally, the change in shape of the voltammogram as a whole may reflect a distribution of diffusion domain sizes: there may be some which are very close together and some which are sparse, and so the summation of voltammetry reflecting different diffusional behaviours distorts the voltammetry.

The correctness of the edge plane activity model for HOPG electrode kinetics is best justified by the close correlation of experiment and two-dimensional simulated voltammetry, using the diffusion domain approximation.

## 6.8 Partially Blocked Electrodes

### Problem

(i) How may the problem of a partially blocked electrode (PBE) be compared to that of a microelectrode array? What conditions are there on the distribution of a blocking adsorbate to ensure that it does not affect experimental voltammetry?

(ii) The diffusion domain approximation was introduced by H. Reller *et al.* [*J. Electroanal. Chem.* **138** (1982) 65] to analyse partially blocked electrodes. Explain how this approximation simplifies the task of simulating such an electrode.

(iii) Almost all electrode surfaces are 'rough' at some length scale. Why is a rough electrode similar to a partially blocked electrode? Why might we expect that roughness on the scale of a few nanometres will not affect the observed cyclic voltammetry?

### Solution

(i) For both an array and a PBE there exist regions of electroactivity separated by regions of insulating material. So long as the separation between neighbouring sites of electroactivity is short compared to the distance which the electroactive species can diffuse on the experimental timescale, the presence of insulator does not constrain the ability of this species to reach a site of electrolysis, and therefore the current is largely unaffected.

This criterion means that the behaviour of a PBE is critically controlled by the *distribution* of the blocking adsorbate. If the blocking adsorbate is concentrated in a macroscale region, the current will be significantly diminished since this region cannot be diffused across during the experiment, and therefore any electroactive species near it at the beginning of the experiment will not be able to react. By contrast, if blocking adsorbate is distributed in regions with dimensions of the order of microns, the electroactive species initially adjacent to it is not prevented from reaction, since it can diffuse towards an electroactive site during the experiment.

(ii) The diffusion domain approximation assumes that the partially blocked electrode can be simulated like an array of microdiscs, and so each electrode has a recognised 'domain' of solution closer to it than any other electrode. Then an extended array can be simulated by recognising that since each domain within the array behaves equivalently, the flux of any species across the boundary between neighbouring domains must be zero.

Therefore, it suffices to model a single domain and to multiply the result by the number of domains on the electrode surface. The simulation is further simplified by the associated approximation that the domain can be treated as a cylinder with equal basal area to the actual domain, so only a two-dimensional 'slice' needs to be simulated, which is much easier than the full three-dimensional problem.

(iii) D. Menshykau *et al.* [*J. Phys. Chem. C* **112** (2008) 14428] discussed the theory of electrode roughness. A rough electrode is comparable to a partially blocked electrode since some regions – wells and troughs – are less accessible to the electroactive species than peaks on the surface. However, so long as the roughness occurs on nanometre length scales, as is typical, the diffusional Case 4 applies at almost all timescales since diffusional distances are large with respect to defect size. Therefore, the cyclic voltammetry is independent of the roughness and only reflects diffusion to the overall electrode area.

# 7

## Cyclic Voltammetry: Coupled Homogeneous Kinetics and Adsorption

## 7.1 EE Mechanism and Comproportionation

### Problem

Consider the following mechanism:

$$A + e^- \rightleftharpoons B \qquad E^{\ominus}_{f,A/B}$$

$$B + e^- \rightleftharpoons C \qquad E^{\ominus}_{f,B/C}$$

$$A + C \overset{k_f}{\rightleftharpoons} 2B \qquad K_{eq}$$

(i) Derive an expression for $K_{eq}$ as a function of the formal potentials $E^{\ominus}_{f,A/B}$ and $E^{\ominus}_{f,B/C}$.

(ii) Assuming $E^{\ominus}_{f,A/B} \gg E^{\ominus}_{f,B/C}$, how will the value of the forward rate constant ($k_f$) for the chemical step affect the measured voltammetric response? Assume that all the diffusion coefficients are equal, the diffusion is linear and semi-infinite and the electron transfer rate is fast (i.e. reversible).

### Solution

(i) We know that

$$K_{eq} = \exp\left[-\frac{\Delta G^{\ominus}}{RT}\right]$$

for the case in question

$$\Delta G^{\ominus} = \Delta G_1^{\ominus} - \Delta G_2^{\ominus}$$

where

$$\Delta G_1^{\ominus} = -FE_{f,A/B}^{\ominus}$$
$$\Delta G_2^{\ominus} = -FE_{f,B/C}^{\ominus}$$

Thus we can write

$$K_{eq} = \exp\left[-\frac{F}{RT}\left(E_{f,B/C}^{\ominus} - E_{f,A/B}^{\ominus}\right)\right]$$

Given that $E_{f,A/B}^{\ominus} \gg E_{f,B/C}^{\ominus}$, this means that the comproportionation step will be highly favourable, since in this case $K_{eq} \gg 1$.

(ii) Under the conditions outlined in the problem the effects of the comproportionation process are not observable in a voltammetric experiment. For a mathematical proof of this point the reader should refer to Problem 4.13.

## 7.2  EE mechanism: The Reduction of $[(\eta^6\text{-}C_6Me_6)_2Ru][BF_4]_2$

### Problem

The reduction of $[(\eta^6-C_6Me_6)_2Ru][BF_4]_2$ is known to be a two-electron process in acetonitrile with 0.5 M $Bu_4NPF_6$ at a platinum electrode. Simulated voltammetry for this system is depicted in Fig. 7.1; the values for this simulation were taken from D.T. Pierce and W.E. Geiger [*J. Am. Chem. Soc.* **111** (1989) 7636]. At low scan rates,

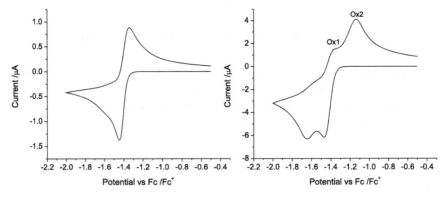

**Fig. 7.1** Simulated voltammetry for the reduction of $(\eta^6-C_6Me_6)_2Ru^{2+}$ at a Pt electrode at low ($0.2\,V\,s^{-1}$) and high ($10\,V\,s^{-1}$) scan rates.

one near-reversible voltammetric feature is observed which corresponds to the two-electron reduction of $(\eta^6-C_6Me_6)_2Ru^{2+}$ to $(\eta^6-C_6Me_6)(\eta^4-C_6Me_6)Ru$. At high scan rates, two peaks are seen in both the forward and reverse scans.

(i) Explain the observed change in the voltammetry. What does this tell us about the structural change associated with the reduction? Write down chemical equations associated with these electron transfers.

(ii) In order to fully describe this system it is necessary to include the disproportionation step. The rate of disproportionation has been measured to have a relatively low value of $6 \times 10^4$ $M^{-1}$ $s^{-1}$. In the absence of the disproportionation step, the oxidation peak Ox1 decreases in height and the oxidation peak Ox2 increases in height; these peaks are indicated in Fig. 7.1. Note that the conditions outlined in Problem 7.1 do not hold here as the electron transfers are not both fully reversible (at high scan rate) and the diffusion coefficients for the oxidised and reduced species differ. Explain this observation and write an equation for the disproportionation mechanism.

(iii) What would the voltammetry look like for this system, if the rate of disproportionation were far larger (i.e. diffusionally controlled, $\sim 10^{10}$ $M^{-1}$ $s^{-1}$).

## Solution

(i) At high scan rates the voltammetric wave for the reduction of the ruthenium complex splits into two peaks. This results from the rate of the second electron transfer being substantially lower than that of the first.

This experiment is significant for two reasons. First, it provides direct evidence that the electron transfer should be viewed as occurring in two discrete steps, and not in a concerted manner.

Second, that the reduction of the monocation species to the neutral species is substantially slower than the first electron transfer strongly implies that the transfer of the second electron results in a large conformational change. It is at this point that one of the arene ligands changes its bonding from $\eta^6$ to $\eta^4$, where the term $\eta$ refers to the hapticity of the ligand. This means that the bonding changes such that four rather than six p-orbitals on carbon atoms are involved in bonding to Ru. Accordingly the heterogeneous electron transfer steps are:

$$(\eta^6-C_6Me_6)_2Ru^{2+} + e^- \rightleftharpoons (\eta^6-C_6Me_6)_2Ru^+$$

$$(\eta^6-C_6Me_6)_2Ru^+ + e^- \rightleftharpoons (\eta^6-C_6Me_6)(\eta^4-C_6Me_6)Ru$$

(ii) In the reverse scan of the voltammogram, the peak Ox1 corresponds to the one-electron oxidation of the monocation, $(\eta^6-C_6Me_6)_2Ru^+$, to the dication,

$(\eta^6-C_6Me_6)_2Ru^{2+}$. Peak Ox2 corresponds to the two-electron oxidation of the neutral species, $(\eta^6-C_6Me_6)(\eta^4-C_6Me_6)Ru$, to the dication. In the absence of a chemical step during the forward scan, all of the electroactive material at the electrode has been reduced to the neutral species, and hence no peak is observed at Ox1. It is not until Ox2 that the oxidation of the neutral species may occur due to the overpotential associated with the slow oxidation of $(\eta^6-C_6Me_6)(\eta^4-C_6Me_6)Ru$ to $(\eta^6-C_6Me_6)_2Ru^+$.

The disproportionation reaction is:

$$2(\eta^6-C_6Me_6)_2Ru^+ \rightleftharpoons (\eta^6-C_6Me_6)_2Ru^{2+}$$
$$+ (\eta^6-C_6Me_6)(\eta^4-C_6Me_6)Ru$$

In the case where this reaction occurs, the concentration of $(\eta^6-C_6Me_6)_2Ru^+$ becomes non-zero, and hence a peak is observed at Ox1.

(iii) Where the disproportionation step is fast, a single voltammetric wave will be observed even at high scan rates. On the forward scan, the monocation formed by the first reduction may undergo rapid disproportionation to form the neutral and dication species. This recovered dication may then undergo further reduction. Overall this results in the process being a two-electron reduction of the dication.

On the reverse scan the neutral species produced is able to undergo comproportionation with the dication species, forming the monocation which is more rapidly reduced at the electrode surface than the neutral species. Again this results in an overall two-electron process.

## 7.3 EC$_2$ Mechanism: The Reduction of the 2,6-Diphenyl Pyrylium Cation

### Problem

Figure 7.2 depicts simulated voltammetry for the reduction of 2,6-diphenyl pyrylium (DPP$^+$) in an acetonitrile solution at various scan rates, ranging from $10^2$ to $10^5$ V s$^{-1}$. This system is described as being an EC$_2$ process.

(i) What do the terms E and C$_2$ stand for?
(ii) Explain the observed change in the voltammetric wave as a function of scan rate. How would you expect its shape to change as a function of the concentration of DPP$^+$?

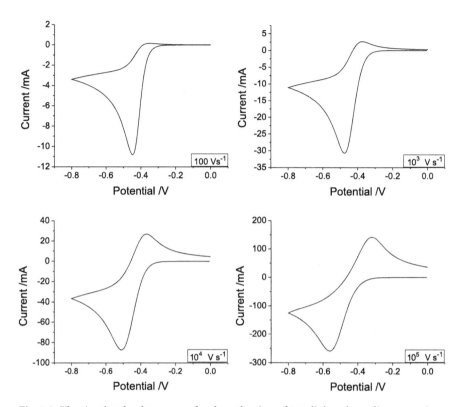

**Fig. 7.2** The simulated voltammetry for the reduction of 2,6-diphenyl pyrylium at various scan rates.

## Solution

(i) The reaction is described as being an $EC_2$ mechanism, which means that it involves an electron transfer step (E) which is followed by a bimolecular chemical step ($C_2$), according to the notation due to A.C. Testa and W. Reinmuth [*Anal. Chem.* **33** (1961) 1320].

(ii) The reduction of $DPP^+$ is known to proceed via the following mechanism:

$$DPP^+ + e^- \rightleftharpoons DPP^\cdot$$

$$2DPP^\cdot \xrightarrow{k} (DPP)_2$$

Using this mechanism we may explain the observed voltammetry. At scan rates which are slow compared to the chemical step ($100 \ V \ s^{-1}$), as the $DPP^+$ is reduced it is able to undergo the chemical step and form the species $(DPP)_2$. As $(DPP)_2$ is not electroactive within the experimental potential window, no peak is observed on the reverse scan leading to an irreversible voltammetric signal.

As the scan rate increases, the chemical process becomes relatively slow on the timescale of the experiment. Consequently, at very high scan rates ($10^5$ V s$^{-1}$) the chemical step is outrun by the voltammetric scan, and a reverse peak is observed. Because the DPP˙ is not consumed by the chemical reaction in the course of the scan, this reverse peak corresponds to the reoxidation of DPP˙ to DPP$^+$.

The simulated data shown are for a 1 mM solution of DPP$^+$. It should be noted that the chemical step is bimolecular in nature. Hence, if the concentration of DPP$^+$ were lowered, this would lead to a proportionally lower rate of formation of (DPP)$_2$ and as such one would expect to observe a peak on the reverse scan at lower scan rates. Conversely, if the concentration of the species were increased, this would lead to an increase in the rate of formation of (DPP)$_2$ and hence one would expect to see an irreversible voltammetric wave at higher scan rates.

## 7.4 Analysis of an Unknown Reaction Mechanism

## Problem

Species A was studied using cyclic voltammetry at a range of scan rates. The first step is known to be a one-electron oxidation:

$$A \rightleftharpoons B^+ + e^-$$

The resulting voltammetry is shown in Fig. 7.3.

(i) How does the scan rate dependence of the voltammetry suggest possible follow-up kinetics?
(ii) On the assumption that the mechanism is EC or EC$_2$, how might the order of the chemical step be determined voltammetrically?
(iii) How might the possible influence of adsorption be discounted?

## Solution

(i) The disappearance of the back peak at low scan rates indicates that the product of the oxidation of A is being consumed by a chemical reaction, such that it is no longer present in solution near to the electrode in order to be reduced in the reverse sweep.

   The consumption of the product also accelerates the initial electron transfer reaction by Le Chatelier's principle, thereby explaining the slight shift of the

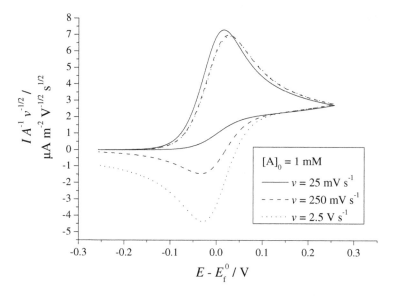

**Fig. 7.3** Appropriately normalised voltammetry for the species A under analysis, showing the influence of follow-up kinetics.

forward peak to a lower potential. The variation of the size of the back peak with scan rate indicates that the chemical kinetics of this chemical step are comparable to the timescale of the voltammetric experiment.

(ii) Because the $EC_2$ process has second-order kinetics in the chemical step, the absolute concentration of the electroactive species will affect the voltammetric response, other than by scaling. In an $EC_2$ process, an increased concentration of A yields an increased concentration of the product B, and hence since the rate of B depletion is proportional to $[B]^2$, this depletion will occur more rapidly at elevated initial concentration, $[A]_0$.

This can be demonstrated by dimensional analysis of the rate equations. Suppose:

$$\frac{\partial [A]}{\partial t} = -k_1 [A]$$

or

$$\frac{\partial [A]}{\partial t} = -k_2 [A]^2$$

Introducing a normalised time $\tau = (F/RT)\,vt$ and a normalised concentration $c = [A]/[A]_0$:

$$\frac{\partial c}{\partial \tau} = -K_1 c$$

or

$$\frac{\partial c}{\partial \tau} = -K_2 c^2$$

where

$$K_1 = k_1 \times \frac{RT}{Fv}$$

$$K_2 = k_2 \times [A]_0 \times \frac{RT}{Fv}$$

thus clarifying that the $EC_2$ voltammetry depends on $[A]_0$ (in a manner other than just scaling the current).

Hence if the mechanism is $EC_2$, the size of the back peak relative to the forward peak will vary with concentration; this is not the case for an EC process.

(iii) Adsorptive processes tend to have somewhat different waveshapes due to the different interplay of exhaustion of electroactive material with increasing rate of reaction at the electrode surface. In particular, $I_{pf} \propto v$ rather than $I_{pf} \propto v^{\frac{1}{2}}$ is characteristic of adsorption of A, as is total collapse of current at high overpotential due to complete exhaustion of the electroactive material. Of course, numerical simulation is the best way to compare different mechanisms in this way, since experimental voltammetry can be compared with a variety of mechanisms and 'best fit' can be used to infer the likely mechanism.


## 7.5  EC Mechanism: Diethyl Maleate

### Problem

Figure 7.4 depicts the simulated voltammetric response for the reduction of diethyl maleate (DEM) at three different scan rates. It should be highlighted that in order to allow direct comparison of the voltammograms, the current has been normalised with respect to the square root of the scan rate. DEM consists of a double bond where the ester groups are situated *cis* to each other. The stereoisomer of DEM is diethyl fumarate (DEF) where the ester groups are situated *trans* to each other, as shown in Fig. 7.5. The reduction potential for DEF is known to be less negative than that of DEM.

(i) Explain the observed shift in the reverse peak position as a function of scan rate.

(ii) Explain the observed increase in the forward peak height and its shift to a less negative potential at low scan rates.

(iii) How many voltammetric peaks would you expect to observe (at $0.1 \text{ V s}^{-1}$) if a second scan was run directly after the first scan? Sketch your answer.

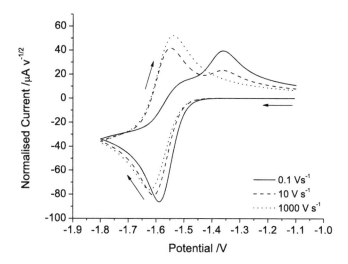

**Fig. 7.4** The simulated cyclic voltammograms at various scan rates for the reduction of DEM at a macroelectrode. The current has been normalised with respect to the square root of scan rate.

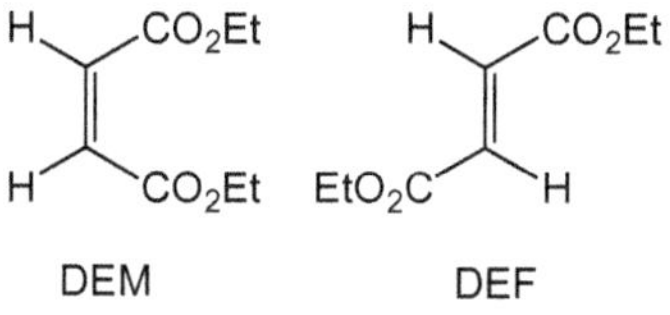

**Fig. 7.5** Chemical structures for DEM and DEF.

## Solution

The reduction of DEM is known to proceed via the following mechanism:

$$DEM + e^- \rightleftarrows DEM^{\cdot-} \qquad E_f^\ominus = -1.58\,V$$

$$DEM^{\cdot-} \xrightarrow{k} DEF^{\cdot-}$$

where the one-electron reduction of DEM is followed by a chemical step in which the radical anion is able to freely rotate around the central carbon–carbon bond, and hence is able to form the less sterically hindered isomer (DEF). The DEF$^{\cdot-}$ that is produced may be reoxidised at a less negative potential:

$$DEF + e^- \rightleftarrows DEF^{\cdot-} \qquad E_f^\ominus = -1.38\,V$$

Using this mechanism we may now explain the observed voltammetry shown in Fig. 7.4.

(i) At high scan rates (e.g. $1000\,\mathrm{V\,s^{-1}}$) the chemical step is relatively slow on the timescale of the experiment, such that only one voltammetric signal is observed. This reversible wave corresponds to the direct reduction of DEM to its radical anion and the reoxidation of this species on the reverse scan.

As the scan rate is lowered, the DEM·⁻ is able to isomerise to form DEF·⁻. The oxidation of DEF·⁻ corresponds to the voltammetric peak at $-1.36\,\mathrm{V}$ on the reverse scan, and hence at the intermediate scan rate ($10\,\mathrm{V\,s^{-1}}$) a mixture of products is formed by partial isomerisation, leading to two peaks being observed on the reverse scan.

At low scan rates ($0.1\,\mathrm{V\,s^{-1}}$) the majority of the DEM·⁻ is converted to its isomer DEF·⁻ before the reverse sweep, and so only one voltammetric feature is observed on the reverse scan, situated at $-1.36\,\mathrm{V}$.

(ii) The influence of the chemical step upon the electrochemical reduction of DEM has both thermodynamic and kinetic implications. Thermodynamically the presence of the chemical step lowers the formal potential of the redox system. This may be exemplified through consideration of the Nernst equation:

$$E = E_f^\ominus - \frac{RT}{F} \ln \frac{[\mathrm{DEM^{\cdot-}}]}{[\mathrm{DEM}]} \tag{7.1}$$

The DEM·⁻ formed is consumed by the chemical step such that, at equilibrium, we can write

$$[\mathrm{DEM^{\cdot-}}] = \frac{[\mathrm{DEF^{\cdot-}}]}{K_{eq}}$$

where $K_{eq}$ is the equilibrium constant for the chemical step. Inclusion of this into the Nernst equation given in Eq. 7.1 gives

$$E = E_f^\ominus - \frac{RT}{F} \ln \frac{[\mathrm{DEF^{\cdot-}}]}{K_{eq}[\mathrm{DEM}]}$$

Inspection of this equation allows us to see that as the equilibrium constant for the chemical step increases the associated equilibrium potential for the redox couple will become less negative, i.e. the species becomes easier to reduce leading to a positive shift in the voltammetric peak position.

The above discussion is based upon the system being at equilibrium. Kinetically speaking, the reduction of DEM is reversible and so the flux measured on the forward scan may be described by the following equation:

$$J = k^0 \exp\left[\frac{-\alpha F}{RT}(E - E_f^\ominus)\right][\mathrm{DEM}]_0$$

$$- k^0 \exp\left[\frac{(1-\alpha)F}{RT}(E - E_f^\ominus)\right][\mathrm{DEM^{\cdot-}}]_0$$

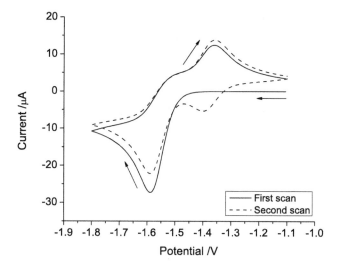

**Fig. 7.6** Simulated first and second cyclic voltammograms for the reduction of DEM at a macroelectrode at $0.1\,\text{V}\,\text{s}^{-1}$.

where $J$ is the flux, $k^0$ is the standard electrochemical rate constant, $E_f^\ominus$ is the formal electrochemical potential for DEM and $[i]_0$ is the concentration of species $i$ at the electrode. Due to $k^0$ being large, both terms will contribute to the net flux. At slower scan rates the concentration of DEM$^{\cdot-}$ is lowered due to the presence of the chemical process (isomerisation). This leads to an increase in the net flux and subsequently the normalised peak current (with respect to square root of scan rate) is increased.

(iii) On the first cycle of the low scan rate voltammogram (at $0.1\,\text{V}\,\text{s}^{-1}$), a significant quantity of DEF has been formed at the electrode surface. Consequently, if a second cycle is performed, one would expect to see two peaks in the forward scan: one due to the reduction of DEF adjacent to the electrode that was generated in the first cycle, and the other from the reduction of further DEM diffusing towards the electrode from bulk solution. Figure 7.6 shows the simulated voltammograms for the first and second cycles for the reduction of DEM at a scan rate of $0.1\,\text{V}\,\text{s}^{-1}$.

## 7.6  ECE Mechanism: *p*-chlorobenzonitrile

## Problem

An ECE mechanism is similar to that in Problem 7.3, except the product of the chemical step is electroactive and is able to undergo a further electron transfer. An example of this is found with the reduction of *p*-chlorobenzonitrile (Cl–$C_6H_4$–CN) which is able to undergo the following mechanism:

$$Cl\text{–}C_6H_4\text{–}CN + e^- \rightleftharpoons Cl\text{–}C_6H_4\text{–}CN^{\cdot-} \qquad E_f^\ominus = -1.96\,V$$

$$Cl\text{–}C_6H_4\text{–}CN^{\cdot-} + H^{\cdot} \rightarrow C_6H_5\text{–}CN + Cl^-$$

$$C_6H_5\text{–}CN + e^- \rightleftharpoons C_6H_5\text{–}CN^{\cdot-} \qquad E_f^\ominus = -2.32\,V$$

where H' is abstracted from the solvent. The chemical step is highly thermodynamically driven and its kinetics are fast.

Sketch the voltammetry for the reduction of Cl–$C_6H_4$–CN at high, medium and low scan rates (relative to the rate of the chemical step).

## Solution

Figure 7.7 shows three simulated voltammograms for the reduction of Cl–$C_6H_4$–CN various scan rates. At a high scan rate (Fig. 7.7(a)), only one reversible voltammetric wave is observed, at $\approx -1.96\,V$. This corresponds to the reduction and reoxidation of the Cl–$C_6H_4$–CN species. Due to the high scan rate, the rate of the chemical step is slow on the timescale of the experiment, such that none of the Cl–$C_6H_4$–CN$^{\cdot-}$ formed on the forward scan reacts via the chemical process to form $C_6H_5$–CN.

In contrast, at relatively low scan rates (Fig. 7.7(c)) the forward reductive scan exhibits two reductive peaks situated at $\approx -1.96\,V$ and $\approx -2.32\,V$. The first voltammetric wave corresponds to the reduction of the Cl–$C_6H_4$–CN species to Cl–$C_6H_4$–CN$^{\cdot-}$. Due to the relatively low scan rate this species is able to further react within the voltammetric scan, to form $C_6H_5$–CN. Hence the $C_6H_5$–CN now present adjacent to the electrode may be further reduced to $C_6H_5$–CN$^{\cdot-}$ at $\approx -2.32\,V$. On the reverse scan the $C_6H_5$–CN$^{\cdot-}$ is reoxidised leading to a peak.

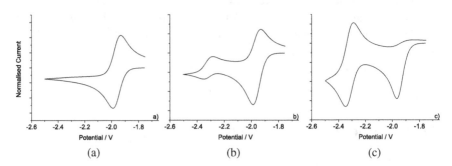

**Fig. 7.7** Simulated voltammetry for the reduction of Cl–$C_6H_4$–CN at (a) fast (b) medium and (c) low scan rates, where the scan rates have been described relative to the rate of the chemical step.

No peak is observed on the reverse scan for the oxidation of $Cl-C_6H_4-CN^{\cdot-}$. This is because the chemical step is both irreversible and fast (on the timescale of the experiment), and hence any $Cl-C_6H_4-CN^{\cdot-}$ present has been consumed to form $C_6H_5-CN$.

At intermediate scan rates (Fig. 7.7(b)), both redox couples are observed. This arises due to the fact that the chemical step is unable to go to completion over the timescale of the experiment. Hence after the forward scan both $C_6H_5-CN^{\cdot-}$ and $Cl-C_6H_4-CN^{\cdot-}$ are present in solution close to the electrode.

## 7.7 ECE vs DISP 1: Voltammetry of Fluorescein

## Problem

The ECE (Eq. 7.2) and DISP 1 (Eq. 7.3) mechanisms for the reduction of fluorescein (F) are as follows [R.G. Compton *et al.*, J. Chem. Soc. Faraday Trans. 1, **84** (1988) 2057]:

$$F + e^- \rightleftarrows F^{\cdot-}$$
$$F^{\cdot-} + H^+ \rightarrow FH^{\cdot} \tag{7.2}$$
$$FH^{\cdot} + e^- \rightleftarrows FH^-$$

$$F + e^- \rightleftarrows F^{\cdot-}$$
$$F^{\cdot-} + H^+ \rightarrow FH^{\cdot} \tag{7.3}$$
$$F^{\cdot-} + FH^{\cdot} \rightleftarrows F + FH^-$$

(i) Given that $E_f^{\ominus}$ is more negative for the reduction of F than for the reduction of $FH^{\cdot}$, justify that the disproportionation reaction in the DISP 1 mechanism is thermodynamically favoured.

(ii) Explain qualitatively why the voltammetry of these two processes is almost equivalent, if the protonation step is rate-determining.

(iii) Double potential step chronoamperometry has been employed to explore the mechanism. The first step is to a potential sufficient to drive the two-electron reaction, and the second step was to a potential intermediate between the two reduction potentials.

Explain why double potential step chronoamperometry is able to distinguish between the two mechanisms. Include a sketch of the concentration profiles immediately before the second step.

## Solution

(i) We can write the homogeneous reaction as the sum of the oxidation of $F^{\cdot-}$ to F (denoted as reduction (1) below), and the reduction of $FH^{\cdot}$ to $FH^-$

(denoted as reduction (2) below).

$$\Delta G^\ominus = -\Delta G_1^\ominus + \Delta G_2^\ominus = nFE_{f,1}^\ominus - nFE_{f,2}^\ominus$$
$$= F(E_{f,1}^\ominus - E_{f,2}^\ominus)$$

If $E_{f,1}^\ominus < E_{f,2}^\ominus$, $\Delta G^\ominus$ is negative, and so the reaction is thermodynamically favoured.

(ii) In both mechanisms, the rate-determining step is the generation of FH$^\cdot$ by the protonation of the fluorescein radical anion. This protonation will occur at a rate proportional to the concentration of the radical anion and, assuming a roughly uniform proton concentration in the solution, will take place across the developing diffusion layer of the radical anion. Therefore, the voltammetry in either case will depend on the rate at which FH$^\cdot$ is generated, since this species is rapidly reduced once present in solution close to the electrode surface.

In either the ECE or DISP 1 case, the generation of FH$^\cdot$ is rate-limiting. The current is therefore controlled by the transport of F to the surface and its protonation, but not by the exact mechanism by which the second electron is transferred. This is because in a potential sweep experiment it is impossible to probe the potential region between the reduction potential of FH$^\cdot$ and the reduction potential of F, under conditions where FH$^\cdot$ is present in solution.

(iii) This latter difficulty can be resolved using a chronoamperometric method.

If FH$^\cdot$ reacts principally by heterogeneous electron transfer at the electrode surface (ECE), its concentration will be depleted there, but depending on the rate at which it is generated, some of the radical anion may have diffused away from the electrode *before* protonation. Therefore, FH$^\cdot$ is also generated further from the electrode and will diffuse back towards it to react.

By comparison, if the disproportionation reaction between F$^{\cdot-}$ and FH$^\cdot$ is fast (DISP 1), the radical FH$^\cdot$ will be consumed as soon as generated whether close to the electrode or not, since the presence of F$^{\cdot-}$ is a pre-condition for its formation, and therefore it will rapidly disproportionate with a further molecule once formed.

Consequently, at some short time after a potential step to a potential where the reduction of fluorescein is fast, there is likely to be some FH$^\cdot$ present in solution if the disproportionation reaction is inactive (ECE), but not if this reaction is fast (DISP 1). On a further potential step to an intermediate potential where the reduction of FH$^\cdot$ is still driven, but now the oxidation of F$^{\cdot-}$ back to F is driven, the current will be purely oxidative in the absence of FH$^\cdot$, but will be compensated by some reductive current if this species is present.

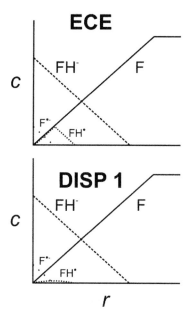

**Fig. 7.8** Schematic concentration profiles following a potential step to a potential where the full conversion from F to FH⁻ is driven, under an ECE (top) and a DISP 1 (bottom) mechanism.

Sketches shown at Fig. 7.8 show how the radical intermediates are retained in ECE since the overall F to $FH^-$ transition only occurs at the electrode, whereas in DISP 1, $FH^.$ is depleted everywhere by its fast reaction with $F^{.-}$.

This allows the cases to be distinguished as the absolute currents will be more negative in the DISP 1 case. In practice, it is sensible to use a simulation program to compare experimental data with the simulated chronoampero-gram to assess whether the difference is significant to within experimental error.

## 7.8 Reduction of Anthracene in DMF

### Problem

The reduction of anthracene (abbreviated A, $C_{14}H_{10}$, Fig. 7.9) in the aprotic solvent dimethylformamide (DMF) has been studied by B.S. Jensen *et al.* [*J. Am. Chem. Soc.* **97** (1975) 5211]. The comparative voltammetry in the presence of differing concentrations of phenol has been discussed by J.-M. Savéant in *Elements of Molecular and Biomolecular Electrochemistry* [(2006) John Wiley and Sons]. Exemplar voltammetry is given in Fig. 7.10.

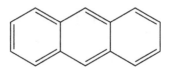

**Fig. 7.9** Anthracene.

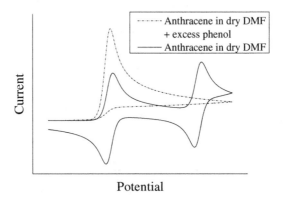

Potential

**Fig. 7.10** Cyclic voltammetry of anthracene in dry DMF in the presence and absence of excess phenol. Note that current and potential are plotted on a negative axis, i.e. the forward wave is cathodic.

(i) Discuss why the presence of phenol in an otherwise aprotic solvent can cause such drastic changes in the voltammetry.
(ii) C. Amatore *et al.* [*J. Electroanal. Chem.* **107** (1980) 353] explored a possible DISP 1 mechanism in the presence of phenol using the double potential step chronoamperometry method discussed in Problem 7.7. What thermodynamically favoured disproportionation reaction might compete with the direct reduction of AH˙?
(iii) Why would one expect the formal reduction potentials of A and AH˙ to be ordered such that this disproportionation is favourable?

## Solution

(i) In an aprotic solvent, the reaction mechanism is straightforward:

$$A + e^- \overset{E_{f,1}^{\ominus}}{\rightleftharpoons} A^{\cdot-}$$

$$A^{\cdot-} + e^- \overset{E_{f,2}^{\ominus}}{\rightleftharpoons} A^{2-}$$

where A is anthracene, $A^{\cdot-}$ is the anthracene radical anion and $A^{2-}$ is the anthracene dianion. Each reduction is successively less favourable, such that $E^{\ominus}_{f,2} \ll E^{\ominus}_{f,1}$. Therefore, two widely separated reversible electrochemical waves are observed, as expected for an EE mechanism. It is possible that the favourable comproportionation reaction between A and $A^{2-}$ may contribute, although this would not be observable in the voltammetry, as discussed in Problem 4.13.

The addition of phenol provides a source of protons and therefore disrupts the EE mechanism by introducing possible irreversible follow-up kinetics. In particular, the anthracene radical anion is prone to protonation:

$$A^{\cdot-} + PhOH \rightarrow AH^{\cdot} + PhO^-$$

The resulting AH$^{\cdot}$ radical is *more* easily reduced than anthracene, and therefore is rapidly reduced and protonated again to yield $AH_2$. The initial protonation is rate-determining for the overall two-electron transfer, and so both electron transfers are observed in a single voltammetric wave, which may be shifted to more positive potentials by accelerating the protonation process, e.g. by elevating the concentration of phenol.

Because the product $AH_2$ (Fig. 7.11) is neither acidic nor electroactive in the potential range studied, its formation is electrochemically irreversible. In the presence of phenol, no peaks are observed in the reverse sweep of the voltammogram, indicating that electroactive product materials such as $A^{\cdot-}$ are fully depleted by the coupled homogeneous chemistry.

(ii) The protonation of the radical anion $A^{\cdot-}$ generates a radical AH$^{\cdot}$ which can undergo disproportionation with that anion:

$$A^{\cdot-} + AH^{\cdot} \rightleftharpoons A + AH^{\cdot-}$$

As discussed in Problem 7.7, conventional voltammetry cannot distinguish this reaction from the case where AH$^{\cdot}$ reacts heterogeneously to gain an electron at the electrode surface, followed by further protonation to form $AH_2$. The disproportionation is thermodynamically favoured if $E^{\ominus}_{f,A} < E^{\ominus}_{f,AH^{\cdot}}$.

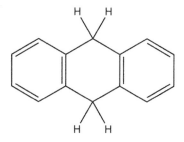

**Fig. 7.11** 9,10-dihydroanthracene.

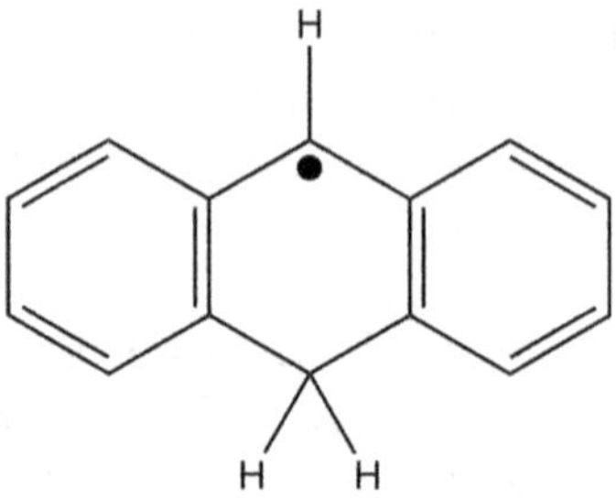

**Fig. 7.12** 9-hydroanthracene radical.

The results from the double potential step chronoamperometry experiments indicated unequivocally that DISP 1 is fast for the case of anthracene in DMF.

(iii) The disproportionation is favourable if the reduction of AH˙ occurs at a higher formal reduction potential (more easily) than the reduction of A. It is clear that this should be the case if we consider that anthracene contains a $14\pi e^-$ system of delocalised electrons, which is aromatic and hence stabilised according to the $4n + 2$ Hückel rule. The addition of an electron causes the loss of aromaticity in the system and hence is strongly thermodynamically disfavoured.

By comparison, the radical species AH˙ already exists in a non-aromatic $13\pi e^-$ form; the resonance structure of two separated $6\pi e^-$ systems (Fig. 7.12) and a localised radical on an $sp^2$ carbon is also relatively stabilised and therefore should predominate. The addition of a further electron to this relatively localised orbital does not significantly alter the aromaticity of the molecule or the general geometry and is therefore much less thermodynamically disfavoured than anthracene reduction, such that its formal potential is more positive.

Hence, the energy gained by oxidising the anthracene radical anion is sufficient to reduce AH˙ and the disporportionation is thermodynamically favoured.

## 7.9 CE Mechanism

## Problem

Figure 7.13 depicts the voltammetric response for the reduction of a species which proceeds by the following mechanism:

$$A \underset{}{\overset{K_{eq}}{\rightleftharpoons}} B$$

$$B + e^- \rightleftharpoons C$$

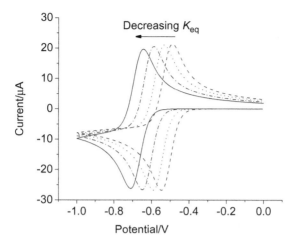

**Fig. 7.13** Simulated voltammetry for a CE mechanism where the value of $K_{eq}$ varies from 1 to $10^{-4}$.

Assume that the electrochemical reduction of B to C is fully reversible, and further that the forward rate constant for the chemical step is very large, to ensure that the species A and B are always at equilibrium on the timescale of the experiment. Explain why the reversible voltammetric wave shifts negatively as the value of $K_{eq}$ is decreased from a value of 1 to $10^{-4}$.

## Solution

The shift in the voltammetric feature is mirroring the change in the equilibrium potential of the reduction of B to C as a function of $K_{eq}$. This may be shown mathematically through consideration of the Nernst equation for the reduction of B to C.

$$E = E_f^\ominus - \frac{RT}{F} \ln \frac{[C]}{[B]}$$

where the concentration of B at equilibrium may be described by

$$[B] = K_{eq}[A]$$

hence

$$E = E_f^\ominus - \frac{RT}{F} \ln \frac{[C]}{K_{eq}[A]} \tag{7.4}$$

From Eq. 7.4, as the value of $K_{eq}$ is decreased, the apparent equilibrium potential for the reduction of B to C becomes more negative.

## 7.10 EC′ Mechanism

## Problem

The following scheme defines an EC′ mechanism,

$$A + e^- \rightleftharpoons B$$

$$B + X \rightleftharpoons A + Y$$

where A undergoes a one-electron reduction to form B, and B may then be reoxidised to A by the species X. As such we may describe the overall system as being the 'electrocatalytic' reduction of X to Y, which is mediated by the A/B redox couple.

With the aid of sketches, describe how the voltammetric response of this system will differ in the presence and absence of an excess of species X.

## Solution

Figure 7.14 depicts the voltammetric response for the reduction of A to B both in the presence and absence of species X. In the absence of the species X, the voltammetric response is that for a reversible one-electron reduction (solid line). In the presence of an excess of species X a large, irreversible voltammetric wave is observed.

The increase in the peak current is due to the reoxidation of B to A by species X. During the forward scan the A/B redox system cycles around such that it is possible for any one molecule to be reduced at the electrode surface multiple times. The

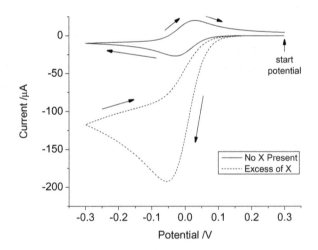

**Fig. 7.14** Simulated voltammetry for an electrocatalytic reduction mechanism, in a potential window from $E - E_f^\ominus = +0.3$ V to $E - E_f^\ominus = -0.3$ V.

lack of a back peak is also explained by the re-oxidation of B to A by species X, so that on the reverse scan the concentration of species B is nearly zero, and thus no oxidation occurs at the electrode.

## 7.11 EC' Mechanism: Cysteine and Ferrocyanide

### Problem

The simulated voltammetry shown in Fig. 7.15 is for the catalytic oxidation of cysteine to cystine at a boron-doped diamond (BDD) electrode, mediated by aqueous ferricyanide ($1 \text{ mM Fe(CN)}_6^{3-}$). The concentration of cysteine has been varied between 0 and 0.5 mM. The mechanism for this system is shown below:

$$\text{Fe(CN)}_6^{4-} \rightleftharpoons \text{Fe(CN)}_6^{3-} + e^-$$

$$\text{Fe(CN)}_6^{3-} + \text{cysteine} \xrightarrow{k_2} \text{Fe(CN)}_6^{4-} + \frac{1}{2}\text{cystine}$$

Numerical values for the simulation have been taken from the work by O. Nekrassova *et al.* [*Electroanalysis* **14** (2002) 1464].

(i) Why does the ferro/ferricyanide redox couple exhibit slow kinetics on the BDD electrode?
(ii) How may this system be used analytically?
(iii) Why, as the concentration of cysteine increases, does the magnitude of the forward peak increase and that of the back peak decrease (as indicated by the arrows on Fig. 7.15)?

### Solution

(i) The electrochemical behaviour of BDD exhibits a number of unusual features. Of prime importance is the fact that the material is a semiconductor, and, consequently, the density of states available for electron transfer is substantially less than that for a metallic electrode. From Marcus theory it is known that the rate of electron transfer is proportional to the density of electronic states in the electrode material [N.S. Lewis *et al.*, *Chem. Phys.* **326** (2006) 15], and hence, in general, far lower rates of electron transfer are observed with BDD electrodes.

Another significant point is that adsorption to BDD is often very weak or even non-existent, and, consequently, any electron transfer which usually involves adsorption (i.e. an inner sphere mechanism) will exhibit low rates of electron transfer. A final point should be made that the BDD surface may be modified to be either hydrogen or oxygen terminated. An example of how

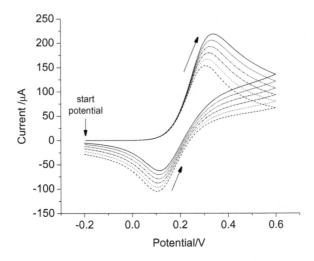

**Fig. 7.15** Simulated voltammetry of the electrocatalytic oxidation of cysteine mediated by ferrocyanide. Arrows indicate how the voltammetry changes as the concentration of cysteine is increased from 0 to 0.5 mM.

this affects the observed voltammetry can be found in the work of C.M.A. Oliveira and A.M. Oliveira Brett [*J. Electroanal. Chem.* **648** (2010) 60].

In regards to ferro/ferricyanide, the oxidation of ferrocyanide is known to be surface sensitive and with the above points in mind it is unsurprising that low rates of electron transfer are observed for the redox couple.

(ii) The forward peak current will scale with the concentration of cysteine present in the solution. A plot of peak current versus cysteine concentration will yield a calibration plot from which it would be possible to determine the unknown concentration of cysteine in a solution. A similar methodology has been successfully applied to the detection of hydrogen sulphide.

(iii) In the absence of cysteine the voltammogram measured is that for ferricyanide in isolation, which can be reduced to ferrocyanide and reoxidised on a reverse scan. As cysteine is added to the solution the oxidative peak increases as the ferricyanide is reduced via the homogeneous chemical step to ferrocyanide, which may then be reoxidised to ferricyanide at the electrode. Hence, this results in a steady increase in the forward peak.

Due to the consumption of the ferricyanide by the homogeneous chemical process, on the reverse scan the concentration of ferricyanide is substantially depleted, and, consequently, the peak measured on the reverse scan is reduced from that expected for the ferro/ferricyanide redox couple in the absence of cysteine.

## 7.12 EC′ Mechanism: Oxygen and Anthraquinone

### Problem

The catalytic reduction of oxygen (1.24 mM) was investigated at a BDD electrode under conditions where the electron transfer is mediated by the reduction of anthraquinone monosulfonate (AQ, 50 $\mu$M). The process is thought to be a $2H^+$, $2e^-$ system, where overall the oxygen is reduced to hydrogen peroxide.

$$AQ + 2e^- + 2H^+ \rightleftharpoons AQH_2$$
$$AQH_2 + O_2 \rightleftharpoons AQ + H_2O_2$$

Figure 7.16 depicts a representative voltammogram for this system.

(i) Why is the direct reduction of oxygen not observed at the BDD electrode?
(ii) On the reverse scan a significant increase in the reductive current is observed at $\approx -0.55$ V. Give a plausible suggestion as to how this feature may occur, in terms of the mechanism.

### Solution

(i) As discussed in Problem 7.11, the rate of electron transfer at a BDD electrode tends to be substantially lower, as compared to other electrodes. This is in part a result of the material being a semi-conductor, and hence there is a far lower density of electronic states in the material than for a metallic electrode.

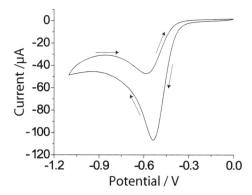

**Fig. 7.16** Experimental voltammetry for the catalytic reduction of oxygen (1.24 mM) at a BDD electrode mediated by anthraquinone monosulfonate (50 $\mu$M). The scan rate is 800 mV s$^{-1}$; arrows indicate scan direction.

The reduction of oxygen is typically irreversible in aqueous solution for a metallic electrode. Accordingly this process is very slow indeed on a BDD electrode, and as such, due to the large overpotential associated with the process, the reduction of oxygen is not observed within the electrochemical window for a BDD electrode.

In contrast, the reduction of anthraquinone monosulfonate is found to be nearly reversible on an Au electrode. Its reduction on a BDD electrode is found to be slower, as expected, and exhibits a larger peak-to-peak separation. For the case in question, this lowering of the rates of electron transfer is useful as it allows us to directly electrochemically probe the rate of reaction between the reduced anthraquinone and oxygen.

(ii) The 'inverse' peak observed on the reverse scan is a highly unusual feature. In order to explain this response it is necessary to note that oxygen is massively in excess of the anthraquinone species, such that the rate-limiting step is the reaction of the reduced anthraquinone with oxygen.

The reduction of anthraquinone involves the transfer of two electrons: although the two transfers are close in potential in aqueous media, the second electron transfer occurs at marginally more negative potential than the first. As a consequence, at low overpotentials significant quantities of the mono-reduced anthraquinone will be formed at the electrode. As the overpotential is further increased the concentration of the monoreduced form decreases as it is reduced futher. On the return scan, as the overpotential is decreased, the concentrations of the monoreduced form will again increase.

It is plausible that monoreduced anthraquinone can drive a one-electron reduction of oxygen to superoxide:

$$AQ + e^- \rightleftharpoons AQ^{\cdot -}$$

$$AQ^{\cdot -} + O_2 \rightarrow AQ + O_2^{\cdot -}$$

Superoxide is highly reactive and will readily disproportionate to form hydrogen peroxide and oxygen. Further, the one-electron reduction of oxygen has a faster rate constant than the two-electron reduction. Hence, it is this effective increase and decrease in the concentration of the monoreduced anthraquinone at the electrode surface, as a function of overpotential, that leads to the observed 'inverse' peak.

## 7.13  Chronoamperometry of Adsorbed Species

## Problem

Derive an equation that describes how the current varies as a function of time in a chronoamperometric experiment with a surface-bound redox species. Hence suggest a suitable plot to assess the magnitude of the electron transfer rate associated with the process.

## Solution

The kinetics of electron transfer for a surface-bound species at a fixed potential essentially follow those of a first-order rate law. Consider the reaction:

$$A + ne^- \rightleftharpoons B$$

where the rate of electron transfer has a value $k$ ($s^{-1}$), the surface coverage ($\Gamma$) is measured in mol cm$^{-2}$ and the total surface coverage is described by

$$\Gamma_{tot} = \Gamma_A + \Gamma_B$$

Since the current, $I$, is the time derivative of charge passed, $Q$, which is proportional to the number of moles reacted

$$I = \frac{dQ}{dt} = nFA \frac{d\Gamma_A}{dt}$$

where $F$ is the Faraday constant and $A$ is the area of the substrate.
  If a first-order rate law is obeyed:

$$\frac{d\Gamma_A}{dt} = -\frac{d\Gamma_B}{dt} = -k\Gamma_A$$

Integrating

$$\ln \Gamma_A = -kt + \ln \Gamma_{tot}$$

$$\Gamma_A(t) = \Gamma_{tot} \exp(-kt)$$

$$I = nFAk\Gamma_{tot} \exp(-kt)$$

on the assumption that at the beginning of the experiment, only A is adsorbed and so $\Gamma_A(t = 0) = \Gamma_{tot}$.
  As such, a plot of $\ln |I|$ vs $t$ should yield a straight line, where the gradient is equal to $-k$.

## 7.14 Voltammetry of an Ideal Adsorbed Species

## Problem

(i) For a linear sweep voltammetric experiment, derive an equation which describes how the current varies as a function of scan rate and the applied potential for the $n$-electron reduction of A to B:

$$A + ne^- \rightleftharpoons B$$

where both A and B are surface-bound.

Assume that:

- the rate of electron transfer is fast enough that a Nernstian distribution of species is maintained at all times
- all adsorption sites are equal
- there are no interactions between the adsorbed species
- the surface activity is equivalent to the surface coverage
- the total surface coverage is independent of the applied potential.

(ii) A typical cyclic voltammogram for an ideal surface-bound species is depicted in Fig. 7.17. Comment on the main features observed.

(iii) From your answer in part (i), derive a parametric expression for the peak height for a surface-bound species. Comment on the major differences between this equation and the Randles–Ševčík equation, which describes the peak height for the voltammetric response of a diffusing species.

## Solution

(i) The assumptions made above allow us to derive the voltammetric response for an ideally adsorbed species. If $\Gamma_i$ is the surface coverage of species $i$, we can write

$$\Gamma_{tot} = \Gamma_A + \Gamma_B \tag{7.5}$$

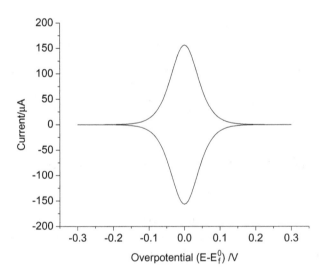

**Fig. 7.17** The cyclic voltammogram of an ideally adsorbed surface species, showing reversible electron transfer kinetics.

The surface coverages are determined by the Nernst equation

$$\frac{\Gamma_A}{\Gamma_B} = \exp(-n\theta) \tag{7.6}$$

where

$$\theta = \frac{F}{RT}(E - E^{\ominus}_{f,A/B})$$

which represents the normalised overpotential applied to the electrode.
Substitution of Eq. 7.5 into Eq. 7.6, followed by rearrangement, gives

$$\Gamma_A = \frac{\exp(n\theta)}{1 + \exp(n\theta)}\Gamma_{tot}$$

The current is given by

$$\frac{I}{nFA} = -\frac{\partial \Gamma_A}{\partial t}$$

where $A$ is the electrode area. The applied potential varies with

$$E = E_{init} - vt$$

noting that we use the negative sign as we are performing a reduction and
hence scanning in a negative direction. Consequently, through differentiation
we get to the required answer

$$\frac{I}{nFA} = \frac{nFv}{RT}\Gamma_{tot}\frac{\exp(-n\theta)}{(1 + \exp(-n\theta))^2} \tag{7.7}$$

Equation 7.7 has been derived for linear sweep voltammetry; the reverse scan
in the cyclic voltammetric experiment is symmetrical about the zero current
axis.

(ii) The main features of the cyclic voltammogram of the surface-bound species
are:

- the forward and backward peaks are symmetrical
- in the reversible limit (as is depicted) the peak-to-peak separation is zero
- at high overpotentials the current is zero due to all of the electroactive
species having been consumed
- the peak width at half peak height is $90.6/n$ mV (where $n$ is the number of
electrons transferred).

(iii) The maximum in Eq. 7.7 occurs when $\theta = 0$, and hence the peak current ($I_{pf}$)
is given by

$$|I_{pf}| = \frac{n^2 F^2 v}{4RT} A \Gamma_{tot}$$

The Randles–Ševčík equation for a reversible diffusional redox species is given by

$$I_{pf} = (2.69 \times 10^5) n^{\frac{3}{2}} A D^{\frac{1}{2}} C^* v^{\frac{1}{2}}$$

where the symbols used are described in Problem 4.6. For the experimental electrochemist these two equations and their dependency on scan rate is of significance. For a surface-bound species a plot of $I_{pf}$ vs scan rate should yield a straight line, whereas in the diffusional case species a plot of $I_{pf}$ vs the square root of scan rate should yield a straight line. Hence, through this analysis it can be readily ascertained whether the electroactive species of interest is surface-bound or diffusional.

## 7.15  Non-Ideal Adsorbed Species

## Problem

The voltammetry of surface-bound species is often found to exhibit non-ideal behaviour, in which the peak width at half peak height is not $90.6/n$ mV. Even under conditions where the peak position is not influenced by the scan rate, the peak-to-peak separation may still be non-zero. A large number of models have been proposed to explain these experimentally observed non-idealities (see M.J. Honeychurch and G.A. Rechnitz, *Electroanalysis* **10** (1998) 285).

(i) Assuming Frumkin type adsorption, derive an equation which describes the peak current for a surface-bound species which may undergo a reversible $n$-electron reduction. Assume that the activity coefficients ($\gamma_i$) for the surface-bound species may be described by

$$\gamma_A = \exp\left(-2a_{AA}\frac{\Gamma_A}{\Gamma_{tot}} - 2a_{AB}\frac{\Gamma_B}{\Gamma_{tot}}\right)$$

$$\gamma_B = \exp\left(-2a_{BB}\frac{\Gamma_B}{\Gamma_{tot}} - 2a_{AB}\frac{\Gamma_A}{\Gamma_{tot}}\right)$$

where $a_{ij}$ is the interaction energy between species $i$ and $j$.

(ii) Use your answer from part (i) to explain how the peak height for the surface-bound species varies with $a$ where

$$a = a_{AA} + a_{AB} - 2a_{AB}$$

For what value of $a$ will the surface-bound species behave ideally?

(iii) Another model for a cause of non-ideality relates to the interfacial potential distribution. Give a brief explanation of the physical origins of interfacial potential distribution.

## Solution

(i) For the equilibrium

$$A + ne^- \rightleftharpoons B$$

the following holds

$$\frac{\gamma_A \Gamma_A}{\gamma_B \Gamma_B} = \exp(-n\theta) \tag{7.8}$$

where the symbols are the same as were used in Problem 7.14, except that now we are no longer assuming that the activity of a surface-bound species is equal to its surface coverage. Note that Eq. 7.8 is the Nernst equation. Through the use of $\Gamma_A + \Gamma_B = \Gamma_{tot}$ we can rearrange Eq. 7.8 as

$$\Gamma_A = \Gamma_{tot} \frac{\exp(n\theta)}{\frac{\gamma_A}{\gamma_B} + \exp(n\theta)}$$

We next use the quotient rule to find the derivative $d\Gamma_A/dt$, remembering that we are doing a reduction and so $E = E_{init} - vt$, as in Problem 7.14:

$$\frac{\partial \Gamma_A}{\partial t} = \Gamma_{tot} \frac{g'(t)h(t) - g(t)h'(t)}{h(x)^2}$$

where

$$g(t) = \exp(n\theta)$$

$$g'(t) = \frac{nFv}{RT} \exp(n\theta)$$

$$h(t) = \frac{\gamma_A}{\gamma_B} + \exp(n\theta)$$

$$h'(t) = \frac{\partial \frac{\gamma_A}{\gamma_B}}{\partial t} + \frac{nFv}{RT} \exp(n\theta)$$

noting that $\frac{\gamma_A}{\gamma_B}$ is time dependent. Now considering the expressions for $\gamma_A$ and $\gamma_B$ and with some straightforward calculation:

$$\frac{\partial \frac{\gamma_A}{\gamma_B}}{\partial t} = -\frac{2a}{\Gamma_{tot}} \cdot \frac{\gamma_A}{\gamma_B} \cdot \frac{\partial \Gamma_A}{\partial t}$$

and rearranging we get

$$\frac{\partial \Gamma_A}{\partial t} = \Gamma_{tot} \frac{g'(t) \frac{\gamma_A}{\gamma_B}}{\left( h(t)^2 - 2a \exp(n\theta) \frac{\gamma_A}{\gamma_B} \right)}$$

from

$$I = nFA \frac{\partial \Gamma_A}{\partial t}$$

and further given that at the peak current, $\exp(n\theta) = 1$ due to the symmetry implied by the Nernst equation, we can write

$$I_{pf} = \frac{n^2 F^2 v}{RT} A \Gamma_{tot} \cdot \frac{\frac{\gamma_A}{\gamma_B}}{\left(\frac{\gamma_A}{\gamma_B} + 1\right)^2 - 2a \frac{\gamma_A}{\gamma_B}}$$

If we assume that $\frac{\gamma_A}{\gamma_B} \approx 1$ then we achieve a simplified expression:

$$I_{pf} = \frac{n^2 F^2 v}{RT} A \Gamma_{tot} \cdot \frac{1}{4 - 2a}$$

$$= I_{pf,ideal} \cdot \frac{2}{2 - a} \tag{7.9}$$

(ii) Where $a < 0$, i.e. the interaction energy between different molecules ($a_{AB}$) is greater than that between like molecules ($a_{AA}$ and $a_{BB}$), the peak current will be lower than that found for the ideal case. Alternatively, when $a > 0$ the peak will be higher than that expected for the ideal case. In cases where $a = 0$ the expression in Eq. 7.9 reduces to that found in Problem 7.14 for a ideal surface-bound species. Figure 7.18 depicts the influence of $a$ upon the observed cyclic voltammetry, demonstrating how the peak height but also peak width varies with $a$.

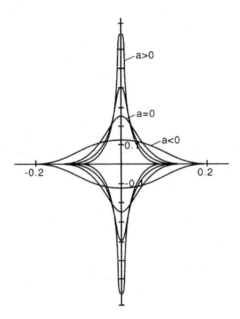

**Fig. 7.18** Voltammetric peaks at 25°C in the case of a Frumkin isotherm. Reproduced from E. Laviron, *J. Electroanal. Chem.* **100** (1979) 263, with permission from Elsevier.

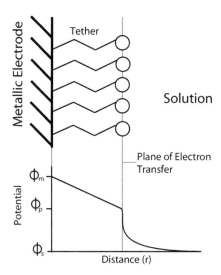

**Fig. 7.19** A diagram showing the change in potential as a function of the distance from the electrode. The molecules of the redox species are situated on the plane of electron transfer.

(iii) A common method for anchoring a redox species to an electrode surface is through the use of a linker group. An excellent example of this is the use of redox-functionalised alkanethiols, which are used with gold electrodes. As a result of these tethering groups, in many experiments the redox species is not situated directly at the electrode interface but is at a finite distance from the surface.

Consequently, there is a drop in potential between the electrode and the redox species as indicated in Fig. 7.19. Here it is assumed that the volume between the electrode and the redox species is not accessible to solvent molecules, and so potential drops approximately linearly across this region. A mathematical model of this effect was first provided by C.P. Smith and H.S. White [*Anal. Chem.* **64** (1992) 2398].

# 7.16 Irreversible Electron Transfer and Adsorbed Redox Species

## Problem

Problems 7.14 and 7.15 have assumed that the coverages of the reduced and oxidised surface-bound species can be described through the use of the Nernst equation. In the case of a surface-bound species where the electron transfer is slow (irreversible), this approximation no longer holds.

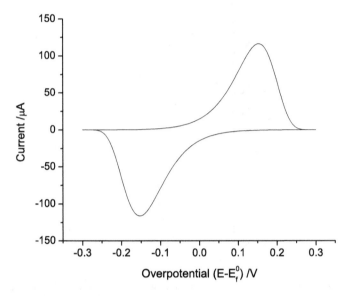

**Fig. 7.20** The cyclic voltammogram of an ideally adsorbed surface species, showing irreversible electron transfer kinetics.

Figure 7.20 depicts the voltammetric response for a surface-bound redox species with irreversible electron transfer kinetics. Comparing this voltammogram with that in Fig. 7.17 we can see a number of important differences, including that the peak-to-peak separation is now non-zero and the peak shape is no longer symmetrical.

(i) Derive an equation which describes the current as a function of the applied potential for the one-electron reduction of a surface-bound species:

$$A_{(ads)} + e^- \rightarrow B_{(ads)}$$

given that the forward rate constant ($k_f$) is

$$k_f = k^0 \exp\left(-\frac{\alpha F}{RT}(E - E_f^{\ominus})\right) \tag{7.10}$$

Note that as we are dealing with an irreversible electron transfer we may ignore the reverse process (i.e. the oxidation of B to A).

(ii) Use your answer to part (i) to derive an equation which describes how the peak position varies as a function of scan rate.

(iii) Using your answer from part (ii), suggest a suitable plot from which you might measure the transfer coefficient ($\alpha$).

(iv) The above parts of this equation have been based on the assumption that the reduction is a one-electron process. How would your answers for parts (i) and (ii) differ for a multi-electron process (see also Problem 2.6)?

## Solution

(i) As in Problem 7.14, we can express the current as

$$I = FA\frac{\partial \Gamma_A}{\partial t} \tag{7.11}$$

But because the Nernst equation does not apply, we cannot express $\Gamma_A$ as a function of $t$ directly. Rather, in order to find the variation in the surface coverage of A with respect to time, it is necessary to solve the following differential equation:

$$\frac{\partial \Gamma_A}{\partial t} = -k_f \Gamma_A \tag{7.12}$$

subject to condition that at $t = 0$ the surface coverage of A is initially $\Gamma_{tot}$.

For a reduction the voltammogram is swept in a negative direction such that $E = E_1 - vt$ and so

$$\int_{\Gamma_{tot}}^{\Gamma_A} \frac{d\Gamma_A'}{\Gamma_A'} = -k^0 \int_0^t \exp\left(-\frac{\alpha F}{RT}(E - E_f^\ominus)\right) dt \tag{7.13}$$

Integration of the right-hand side of this equation is simpler if we make a substitution. Let

$$u = -\frac{\alpha F}{RT}(E - E_f^\ominus)$$

and therefore

$$du = \frac{\alpha Fv}{RT} dt$$

such that we can rewrite Eq. 7.13 as

$$\int_{\Gamma_{tot}}^{\Gamma_A} \frac{d\Gamma_A'}{\Gamma_A'} = -k^0 \frac{RT}{\alpha Fv} \int_{u(0)}^u \exp(u') du'$$

If we assume that the voltammogram is started at a point where no current is being passed, i.e. $\exp(u(0)) = 0$, then

$$\ln \frac{\Gamma_{tot}}{\Gamma_A} = k^0 \frac{RT}{\alpha Fv} \left(\exp\left(-\frac{\alpha F}{RT}(E - E_f^\ominus)\right)\right)$$

$$= \frac{RT}{\alpha Fv} k_f$$

Rearrangement gives

$$\Gamma_A = \Gamma_{tot} \exp\left(-\frac{RT\, k_f}{\alpha Fv}\right) \tag{7.14}$$

It should be noted that as $k_f$ is not constant with time, this imparts a time dependency upon Eq. 7.14 such that at high overpotentials (where $t$ is large) the surface coverage of A will be zero.

Consequently, using Eqs. 7.11, 7.12 and 7.14, the variation of the current as a function of the applied potential is given by

$$I = FAk_f\Gamma_{tot} \exp\left(-\frac{RTk_f}{\alpha Fv}\right) \tag{7.15}$$

where $k_f$ is as defined in Eq. 7.10.

(ii) In order to find the peak position for the current described by Eq. 7.15 we must find the point at which

$$\frac{\partial I}{\partial t} = 0$$

Differentiation of Eq. 7.15 requires the use of product rule where

$$\frac{1}{FA\Gamma_{tot}} \frac{\partial I}{\partial t} = f'(t)g(t) + g'(t)f(t)$$

where according to the relations above

$$f(t) = k_f$$

$$f'(t) = \frac{\alpha Fv}{RT} k_f$$

$$g(t) = \exp\left(-\frac{RTk_f}{\alpha Fv}\right)$$

$$g'(t) = -k_f \exp\left(-\frac{RTk_f}{\alpha Fv}\right)$$

From the above information it clearly follows that when

$$\frac{\partial I}{\partial t} = 0$$

we can write

$$-k_f^2 \exp\left(-\frac{RTk_f}{\alpha Fv}\right) + \frac{\alpha Fv}{RT} k_f \exp\left(-\frac{RTk_f}{\alpha Fv}\right) = 0$$

and therefore

$$k_f = \frac{\alpha Fv}{RT}$$

Expanding this using Eq. 7.10:

$$k^0 \exp\left(-\frac{\alpha F}{RT}(E_{pf} - E_f^{\ominus})\right) = \frac{\alpha F v}{RT}$$

where $E_{pf}$ is the peak potential. Rearrangement gives

$$E_{pf} = E_f^{\ominus} - \frac{RT}{\alpha F} \ln\left(\frac{\alpha F v}{RT k^0}\right) \tag{7.16}$$

This result shows that, for an irreversible surface-bound species, as the scan rate increases the overpotential ($E_{pf}$) of the peak current increases in magnitude (for a reduction as is the case in this question, the value of $E_{pf}$ becomes more negative).

(iii) From inspection of Eq. 7.16 it can be seen that a plot of $E_{pf}$ vs $\ln v$ should yield a straight line of gradient $-RT/\alpha F$. This equation only holds in the irreversible limit, so it is required that the peak-to-peak separation ($\Delta E_{pp}$) is greater than $(200/n)$ mV.

(iv) Where multiple electrons are transferred in a single electrochemical step, then $\alpha$ in the above equations is replaced by $\alpha = (n' + \alpha_{RDS})$ where $n'$ is the number of electrons transferred prior to the rate-determining step and $\alpha_{RDS}$ is a transfer coefficient associated with the rate-determining electron transfer (see also the discussion in Problem 2.6).

## 7.17 Voltammetry of Ferrocyanide/Ferricyanide

### Problem

The anions $[Fe(CN)_6]^{4-}$ (ferrocyanide) and $[Fe(CN)_6]^{3-}$ (ferricyanide) are frequently used redox species.

(i) Of the two species above, which will have the more labile ligands?
(ii) What consequence does this have for their use in electrochemistry?

### Solution

(i) $[Fe(CN)_6]^{4-}$ has Fe in the +2 oxidation state and so has a low spin $d^6$ configuration, whereas $[Fe(CN)_6]^{3-}$ has Fe in the +3 oxidation state and so is low spin $d^5$. Hence, $[Fe(CN)_6]^{4-}$ has the greater ligand field stabilisation energy and as such has less labile $CN^-$ ligands.

Most cyanide salts are highly toxic due to the possibility of the release of cyanide. Considering the above, $[Fe(CN)_6]^{3-}$ is significantly more toxic than its reduced counterpart.

(ii)  The lability of the cyanide ligands is significant for electrochemists due to the ability of ferrocyanide and ferricyanide to complex together to form Prussian blue. Prussian blue is insoluble and forms upon electrode surfaces, causing the electrode to become less electroactive. This problem is circumvented through regular polishing of the electrode.

# 8

---

# Hydrodynamic Electrodes

## 8.1 Channel Electrodes and Limiting Currents

### Problem

Calculate the transport-limited current for the one-electron oxidation of a 1 mM aqueous solution of ferrocyanide, $Fe(CN)_6^{4-}$, at a macro-channel electrode of size 4.0 mm × 4.0 mm in a flow cell of cross-section dimensions 6 mm × 0.4 mm at flow rates ($V_f$) of $10^{-3}$ and $10^{-1}$ $cm^3\,s^{-1}$. Assume a value of $6 \times 10^{-6}$ $cm^2\,s^{-1}$ for the diffusion coefficient of ferrocyanide.

### Solution

The transport-limited current for a one-electron oxidation process at a channel electrode is given by

$$I_{lim} = 0.925\, Fc^*\, wx_e^{\frac{2}{3}} D^{\frac{2}{3}} \left( \frac{V_f}{h^2 d} \right)^{\frac{1}{3}}$$

where $c^*$ is the bulk concentration of ferrocyanide, $w$ is the electrode width (0.4 cm), $x_e$ is the electrode length (0.4 cm), $d$ is the flow cell width (0.6 cm) and $h$ is the half-height of the flow cell (0.02 cm). The other quantities are

$$D = 6 \times 10^{-6}\ cm^2\ s^{-1}$$

$$c^* = 10^{-6}\ mol\ cm^{-3}$$

$$F = 96485\ C\ mol^{-1}$$

in which cm units have been used throughout.

It follows that

$$I_{lim} = 0.925 \times 96485 \times 10^{-6} \times 0.4 \times (0.4)^{\frac{2}{3}}$$

$$\times (6 \times 10^{-6})^{\frac{2}{3}} \left( \frac{V_f}{(0.02)^2 \times 0.6} \right)^{\frac{1}{3}}$$

$$= 1.03 \times 10^{-4} \times V_f^{\frac{1}{3}} \text{ A}$$

So for $V_f = 10^{-3} \text{ cm}^3 \text{ s}^{-1}$, $I_{lim} = 10.3 \,\mu\text{A}$. Similarly, for $V_f = 10^{-1} \text{ cm}^3 \text{ s}^{-1}$, $I_{lim} = 47.8 \,\mu\text{A}$.

## 8.2 Channel Electrodes and Reynolds Number

### Problem

For the channel electrode described in Problem 8.1, calculate for both of the flow rates $10^{-3}$ and $10^{-1} \text{ cm}^3 \text{ s}^{-1}$:
  (i) the linear velocity, $V_0$, at the centre of the channel.
  (ii) the Reynolds number characterising the flow. Comment on the Reynolds number in each case.

### Solution

  (i) The volume solution flow rate, $V_f$ ($\text{cm}^3 \text{ s}^{-1}$) and the centre-line velocity, $V_0$ ($\text{cm s}^{-1}$) are related via the equation,

$$V_f = \frac{4}{3} V_0 hd$$

where $h$ (0.02 cm) is the half-height of the flow cell and $d$ (0.6 cm) is the channel width. Thus

$$V_0 = \frac{3}{4} \times \frac{V_f}{0.02 \times 0.6} \text{ cm s}^{-1}$$

$$= 62.5 \, V_f \text{ cm s}^{-1}$$

Therefore when $V_f$ is $10^{-3} \text{ cm}^3 \text{ s}^{-1}$

$$V_0 = 0.0625 \text{ cm s}^{-1}$$

and when $V_f$ is $10^{-1} \text{ cm}^3 \text{ s}^{-1}$,

$$V_0 = 6.25 \text{ cm s}^{-1}$$

(ii) The dimensionless Reynolds number, Re, characterising the flow is:

$$\text{Re} = \frac{2hV_0}{\nu}$$

where $\nu$ (cm$^2$ s$^{-1}$) is the kinematic viscosity of the solution, which is the ratio of the viscosity to the fluid density. For water at 25°C, $\nu \simeq 10^{-2}$ cm$^2$ s$^{-1}$. Hence for the slower flow

$$\text{Re} = \frac{2 \times 0.02 \times 0.0625}{10^{-2}}$$
$$= 0.25$$

and for the faster flow

$$\text{Re} = \frac{2 \times 0.02 \times 6.25}{10^{-2}}$$
$$= 25$$

The Reynolds number characterises the transition from laminar to turbulent (chaotic) flow. Both Reynolds numbers are sufficiently small that the flow will be laminar in character and not turbulent.

## 8.3  Flow to Rotating Discs and in Channels

### Problem

The transport-limited current at a rotating disc electrode depends on the kinematic viscosity of the solution, whereas the corresponding equation for the channel electrode shows no dependence on this quantity. Comment.

### Solution

The solution flow in a channel electrode under laminar conditions is steady with a parabolic velocity distribution. The streamlines describing the flow at any height in the cell are constant in size, and are always directed along the axis of the cell. In contrast, at a rotating disc the solution is pulled first towards the electrode surface, rotated, and then flung out radially so that the streamlines are changing size and direction as they approach the surface. It is this changing nature of the flow which leads to a dependency of the transport-limited current on the solution kinematic viscosity in the case of the rotating disc, whereas no dependency is seen for the channel electrode.

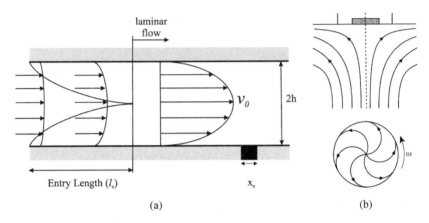

**Fig. 8.1** Schematic diagrams for the solution flow in both a) a channel cell electrode and b) a rotating disc electrode. Reproduced from R.G. Compton *et al.*, *Understanding Voltammetry*, 2nd ed., with permission from Imperial College Press.

## 8.4 Channel Electrodes and ECE Processes

### Problem

The reduction of *m*-iodonitrobenzene in acetonitrile solution at a platinum channel electrode is thought to follow an ECE mechanism [L. Nei *et al.*, *Electroanalysis* 8 (1996) 214]:

$$m\text{-}IC_6H_4NO_2 + e^- \rightarrow m\text{-}IC_6H_4NO_2^{-}$$

$$m\text{-}IC_6H_4NO_2^{-} \xrightarrow{k} I^- + \cdot C_6H_4NO_2$$

$$\cdot C_6H_4NO_2 + HS \xrightarrow{fast} C_6H_5NO_2 + \cdot S$$

$$C_6H_5NO_2 + e^- \rightarrow C_6H_5NO_2^{\cdot-}$$

where HS refers to any H atom available for radical abstraction in the solvent-supporting electrolyte system. The reduction was studied using a channel cell of the following geometry: cell height, $2h = 0.077$ cm, cell width, $d = 0.60$ cm, electrode length, $x_e = 0.40$ cm and electrode width, $w = 0.405$ cm. The following flow rate/$n_{eff}$ data were obtained:

| $V_f$/cm$^3$ s$^{-1}$ | $n_{eff}$ |
|---|---|
| $2.3 \times 10^{-3}$ | 1.41 |
| $3.7 \times 10^{-3}$ | 1.35 |
| $5.2 \times 10^{-3}$ | 1.31 |
| $6.1 \times 10^{-3}$ | 1.27 |
| $6.8 \times 10^{-3}$ | 1.25 |
| $1.1 \times 10^{-2}$ | 1.23 |
| $1.5 \times 10^{-2}$ | 1.11 |

where $n_{eff}$ is the effective number of electrons transferred in the electrode process. The following equations [J.A. Cooper and R.G. Compton, *Electroanalysis* **10** (1998) 141] show how $n_{eff}$ depends on the dimensionless rate constant $K$ where

$$K = k \left( \frac{4x_e^2 h^4 d^2}{9DV_f^2} \right)^{\frac{1}{3}}$$ (8.1)

$$K < 0.59 \quad n_{eff} = 1 + 0.552K - 0.309K^2 + 0.150K^3 \ldots$$ (8.2)

$$0.59 < K < 3.96 \quad n_{eff} = 1.358 + 0.483 \log_{10} K \ldots$$ (8.3)

$$K > 3.96 \quad n_{eff} = 2 - 0.736K^{-1/2} + 0.0613K^{-2} \ldots$$ (8.4)

By constructing a suitable 'working curve', find a value for the rate constant $(s^{-1})$ for the loss of iodide from the radical anion of *m*-iodonitrobenzene. You may assume a value for the mean diffusion coefficient, $D$, of $2.1 \times 10^{-5}$ cm$^2$ s$^{-1}$.

## Solution

Equations 8.2, 8.3 and 8.4 can be used to construct a working curve showing $n_{eff}$ as a function of $K$. This is sigmoidal in shape varying from $n_{eff} = 1$ for small $K$ to $n_{eff} = 2$ for large $K$. The working curve is shown in Fig. 8.2. The working curve

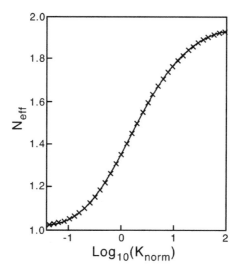

**Fig. 8.2** Working curve for $n_{eff}$ versus $K$ for an ECE reaction. Reproduced from R.G. Compton *et al.*, *Electroanalysis* **8** (1996) 214, with permission from Wiley.

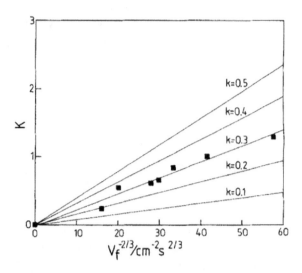

**Fig. 8.3** Analysis of channel electrode current-flow rate data meaured for the reduction of *m*-iodonitrobenzene in terms of an ECE mechanism. Reproduced from R.G. Compton *et al.*, *Electroanalysis* **8** (1996) 214, with permission from Wiley.

can then be used to infer the following values:

| $n_{eff}$ | $K$ |
|-----------|------|
| 1.41 | 1.29 |
| 1.35 | 1.01 |
| 1.31 | 0.84 |
| 1.27 | 0.66 |
| 1.25 | 0.61 |
| 1.23 | 0.54 |
| 1.11 | 0.23 |

Equation 8.1 shows that, if an ECE mechanism operates, a plot of $K$ vs. $V_f^{-2/3}$ will be a straight line through the origin of gradient equal to $k(4h^4 x_e^2 d^2/9D)^{\frac{1}{3}}$. Figure 8.3 shows the plot and that $k \simeq 0.30$ s$^{-1}$.

## 8.5  Channel Electrodes and ECE Processes

## Problem

Show that for a simple ECE reaction:

$$A + e^- \rightleftharpoons B$$
$$B \xrightarrow{k} C$$
$$C + e^- \rightleftharpoons D$$

occurring at a macro-channel electrode constructed such that the Lévêque approximation may be used, the effective number of electrons transferred, $n_{eff}$, in steady-state voltammetry depends on a simple parameter $K$ which is a dimensionless rate constant. Identify any approximations made.

## Solution

For a macro-channel electrode, axial diffusion may be ignored so that the steady-state mass transport equations for species A, B and C are:

$$D\frac{\partial^2[A]}{\partial y^2} - v_x\frac{\partial[A]}{\partial x} = 0$$

$$D\frac{\partial^2[B]}{\partial y^2} - v_x\frac{\partial[B]}{\partial x} - k[B] = 0$$

$$D\frac{\partial^2[C]}{\partial y^2} - v_x\frac{\partial[C]}{\partial x} + k[B] = 0$$

where we have assumed that the diffusion coefficients of A, B and C are equal and take the value $D$. Additionally, the chemical step has been assumed to be irreversible. $x$ is a coordinate parallel to the cell axis, $y$ is a coordinate normal to the electrode surface (see Fig. 8.4) and $v_x$ is the solution velocity in the $x$-direction. We make the Lévêque approximation,

$$v_x = \frac{2v_0 y}{h}$$

where $v_0$ is the centre line velocity (cm s$^{-1}$) and $h$ is the half-height of the flow cell (see Fig. 8.4). We also introduce the dimensionless coordinates,

$$\chi = \frac{x}{x_e}$$

$$\xi = \left(\frac{2v_0}{hDx_e}\right)^{\frac{1}{3}} y$$

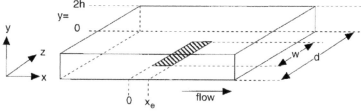

**Fig. 8.4** A typical channel flow cell, with the conventional $x$, $y$ and $z$ axes indicated. Reproduced from N.V. Rees *et al.*, *J. Phys. Chem.* **99** (1995) 7096, with permission from the American Chemical Society.

and so

$$\frac{\partial^2 [A]}{\partial \xi^2} - \xi \frac{\partial [A]}{\partial \chi} = 0$$

$$\frac{\partial^2 [B]}{\partial \xi^2} - \xi \frac{\partial [B]}{\partial \chi} - k \left( \frac{h^2 x_e^2}{4 v_0^2 D} \right)^{\frac{1}{3}} [B] = 0$$

$$\frac{\partial^2 [C]}{\partial \xi^2} - \xi \frac{\partial [C]}{\partial \chi} + k \left( \frac{h^2 x_e^2}{4 v_0^2 D} \right)^{\frac{1}{3}} [B] = 0$$

We introduce the dimensionless rate constant,

$$K = k \left( \frac{h^2 x_e^2}{4 v_0^2 D} \right)^{\frac{1}{3}}$$

so that

$$\frac{\partial^2 [B]}{\partial \xi^2} - \xi \frac{\partial [B]}{\partial \chi} - K[B] = 0$$

and

$$\frac{\partial^2 [C]}{\partial \xi^2} - \xi \frac{\partial [C]}{\partial \chi} + K[B] = 0$$

Inspection of the above equations suggests that the concentration profiles of A, B and C in $\chi$ and $\xi$ are a function of $K$. The effective number of electrons, $n_{\text{eff}}$, is defined as follows:

$$n_{\text{eff}} = \frac{Fw \left[ \int_0^{x_e} D \left. \frac{\partial [A]}{\partial y} \right|_{y=0} + D \left. \frac{\partial [C]}{\partial y} \right|_{y=0} dx \right]}{Fw \int_0^{x_e} D \left. \frac{\partial [A]}{\partial y} \right|_{y=0} dx}$$

where $w$ is the electrode width and $x_e$ is the electrode length (see Fig. 8.4). Alternatively we can write

$$n_{\text{eff}} = \frac{\int_0^1 \left. \frac{\partial [A]}{\partial \xi} \right|_{\xi=0} + \left. \frac{\partial [C]}{\partial \xi} \right|_{\xi=0} \partial \chi}{\int_0^1 \left. \frac{\partial [A]}{\partial \xi} \right|_{\xi=0} dx}$$

$$= n_{\text{eff}}(K)$$

(8.5)

Since in Eq. 8.5, $\xi$ is put equal to zero (the electrode surface) and $\chi$ is integrated over the electrode length (from $\chi = 0$ to $\chi = 1$), $n_{\text{eff}}$ is solely a function of $K$. This assumes that sufficient overpotential is applied that the reaction at the electrode is mass transport controlled.

It is therefore possible to summarise the response of a channel electrode of arbitrary geometry at any flow rate by a working curve which plots theoretical values of $n_{eff}$ as a function of the normalised rate constant $K$.

## 8.6 Channel Electrodes and Entry Length

### Problem

For the channel flow cell and flow rates in Problems 8.1 and 8.2, estimate the distance required for full parabolic flow to be established from the point of entry within the cell. Comment on any implications.

### Solution

For a suitably small Reynolds number, Re, to establish laminar flow, the transition from plug flow (on entry) to parabolic flow is established over an entry length of

$$l_e \simeq 0.1 h \, \text{Re}$$

For the cell in question, $h = 0.02$ cm and the two flow rates of $10^{-3}$ and $10^{-1}$ cm$^3$ s$^{-1}$ correspond (see Problem 8.2) to Reynolds numbers of 0.25 and 25, respectively, so that

$$l_e = 5 \times 10^{-4} \, \text{cm} \quad (V_f = 10^{-3} \, \text{cm}^3 \, \text{s}^{-1})$$

and

$$l_e = 0.05 \, \text{cm} \quad (V_f = 10^{-1} \, \text{cm}^3 \, \text{s}^{-1})$$

In experimental practice, flow rates are usually unlikely to exceed 1 cm$^3$ s$^{-1}$, so that in designing a practical chemical flow cell, an entry length upstream of the working electrode of approximately 0.5 cm should ensure that the appropriate flow profile is established in the vicinity of the channel electrode.

## 8.7 Channel Electrodes and Diffusion Coefficients

### Problem

The oxidation of 1.15 mM ferrocene in the solvent dimethylformamide (DMF) has been studied using a fast flow channel electrode [N.V. Rees *et al.*, *J. Phys. Chem.* **99** (1995) 7096]:

$$Cp_2Fe - e^- \rightleftharpoons Cp_2Fe^+$$

Data for the transport-limited current ($I_{lim}/\mu A$) as a function of flow rate ($V_f/cm^3\ s^{-1}$) were obtained, as shown in Fig. 8.5. Use the Levich equation to determine the diffusion coefficient of ferrocene in DMF given that the cell geometry (Fig. 8.4) was $d = 0.20$ cm, $w = 0.20$ cm, $2h = 1.16 \times 10^{-2}$ cm and $x_e = 12\ \mu m$.

## Solution

The Levich equation predicts that the transport-limited current is given by:

$$I_{lim} = 0.925 F [Cp_2 Fe]_{bulk}\ x_e^{\frac{2}{3}} D^{\frac{2}{3}} w \left(\frac{V_f}{h^2 d}\right)^{\frac{1}{3}}$$

The data shown in Fig. 8.5 are consistent with this equation, in that $I_{lim}$ scales linearly with $V_f^{\frac{1}{3}}$. The gradient of the plot measured from the figure is approximately $6 \times 10^{-6}\ A\ cm^{-1}\ s^{\frac{1}{3}}$. Since

$$F = 96485\ C\ mol^{-1}$$

$$[Cp_2 Fe] = 1.15 \times 10^{-6}\ mol\ cm^{-3}$$

$$x_e = 12 \times 10^{-4}\ cm$$

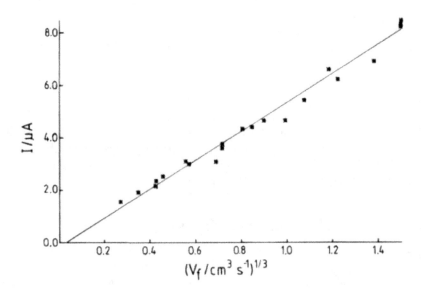

**Fig. 8.5** Transport-limited current ($I_{lim}/\mu A$) versus flow rate ($V_f/cm^3\ s^{-1}$) data obtained from the oxidation of ferrocene (1.15 mM). Reproduced from N.V. Rees *et al.*, *J. Phys. Chem.* **99** (1995) 7096, with permission from the American Chemical Society.

and $d$, $2h$ and $w$ are as given in the question:

$$6 \times 10^{-6} = 0.925 \times 96485 \times 1.15 \times 10^{-6} \times 0.20$$

$$\times \left( \frac{12 \times 10^{-4}}{0.58 \times 10^{-2}} \right)^{\frac{2}{3}} \times \frac{D^{\frac{2}{3}}}{(0.20)^{\frac{1}{3}}}$$

for which

$$D = 1.08 \times 10^{-5} \text{ cm}^2 \text{ s}^{-1}$$

## 8.8 Channel Electrodes and Current Distribution

### Problem

The Levich equation for the transport-limited current at a channel electrode predicts

$$I_{\lim} \propto x_e^{\frac{2}{3}} V_f^{\frac{1}{3}}$$

where $x_e$ is the electrode length and $V_f$ is the solution flow rate (volume per second). How does the current density vary with distance $(x)$ along the electrode? What is the current density at $x = 0$?

### Solution

Suppose that the diffusive flux, $j(0)$, varies with $x$, the distance downstream from the upstream edge of the electrode, as follows:

$$j(x) \propto x^n$$

It follows that

$$I_{\lim} \propto \int_0^{x_e} j(x) \, dx$$

$$\propto \left[ \frac{x^{n+1}}{n+1} \right]_0^{x_e}$$

$$\propto x_e^{n+1}$$

Since the Levich equation shows

$$I_{\lim} \propto x_e^{\frac{2}{3}}$$

it follows that

$$j(x) \propto x^{-\frac{1}{3}}$$

Thus the current density (flux) is predicted to be infinite at the upstream edge of the electrode. In reality, this would not be attained, since finite electrode kinetics preclude the passage of an infinite current. Nevertheless, the flux would be very large at $x = 0$.

## 8.9 Wall-Jet Electrodes and Current Distribution

### Problem

The transport-limited current, $I_{lim}$, at a wall-jet electrode of radius $r_e$ shows the following dependency:

$$I_{lim} \propto r_e^{\frac{3}{4}} V_f^{\frac{3}{4}}$$

where $V_f$ is the volume flow rate. How does the current density (or flux) vary radially over the surface of the electrode?

### Solution

We write $j(r)$ as the the flux normal to the electrode surface at the radial coordinate $r$. Since the total limiting current scales with $r_e^{\frac{3}{4}}$ it follows that

$$I_{lim} \propto \int_0^{r_e} 2\pi r\, j(r)\, dr \propto r_e^{\frac{3}{4}}$$

from which it can be seen that

$$r\, j(r) \propto r^{-\frac{1}{4}}$$

$$j(r) \propto r^{-\frac{5}{4}}$$

showing that the flux is infinite at the centre of the electrode (although an actual infinite flux is precluded by finite electrode kinetics) and that the wall-jet is highly non-uniformly accessible. For a full derivation of the expression for $I_{lim}$ see the work of W.J. Albery and C.M.A. Brett [*J. Electroanal. Chem.* **148** (1983) 201].

## 8.10 Wall-Jet Electrodes and Diffusion Coefficients

### Problem

The one-electron reduction of a 1 mM solution of benzoquinone (BQ) in acetonitrile containing 0.5 M tetrabutylammonium perchlorate as supporting electrolyte,

$$BQ + e^- \rightleftharpoons BQ^{\cdot-}$$

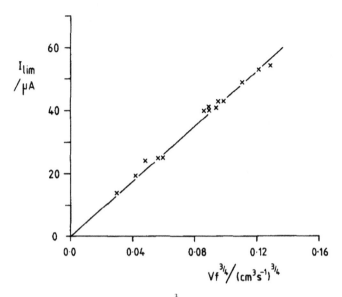

**Fig. 8.6** The linear dependence of $I_{\lim}$ on $V_f^{\frac{3}{4}}$ for the reduction of BQ in acetonitrile at a mercury working electrode in a wall-jet cell. Reproduced from R.G. Compton *et al.*, *J. Electroanal. Chem.* **277** (1990) 83, with permission from Elsevier.

was studied at a wall-jet electrode and the limiting current found to scale with the volume flow rate to the power (3/4)

$$I_{\lim} \propto V_f^{\frac{3}{4}}$$

as shown in Fig. 8.6. The wall-jet had the following geometry: electrode radius, $r_e = 0.075$ cm; nozzle diameter, $a = 0.09$ cm. The kinematic viscosity, $v$, of acetonitrile at the temperature of the experiment (25°C) is $4.425 \times 10^{-3}$ cm$^2$ s$^{-1}$. Estimate the diffusion coefficient ($D$) of benzoquinone in acetonitrile.

## Solution

The transport-limited current for a one-electron reduction is given by

$$I_{\lim} = 1.35 F D^{\frac{2}{3}} v^{-\frac{5}{12}} a^{-\frac{1}{2}} r_e^{\frac{3}{4}} V_f^{\frac{3}{4}} [BQ]_{\text{bulk}}$$

The gradient of the graph in Fig. 8.6 is $4.3 \times 10^{-4}$ A cm$^{-\frac{9}{4}}$ s$^{\frac{3}{4}}$. Hence, substituting values for $F$ (96485 C mol$^{-1}$), $v$, $a$, $r_e$ and $[BQ]_{\text{bulk}}$ ($10^{-6}$ mol cm$^{-3}$) gives

$$D = 1.9 \times 10^{-5} \text{ cm}^2 \text{ s}^{-1}$$

## 8.11 Wall-Jet Electrode and a DISP 1 Process

### Problem

The reduction of 1 mM fluorescein in aqueous solution has been studied at a wall-jet electrode. At pH 13 a simple one-electron reduction is known to occur:

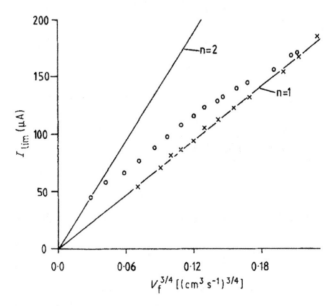

Figure 8.7 shows that the transport-limited current at this pH varies directly with the volume flow rate to the power 3/4 ($V_f^{\frac{3}{4}}$), as expected for an electrode process uncomplicated by coupled homogeneous kinetics. The wall-jet electrode had a radius ($r_e$) of 0.403 cm and a nozzle diameter ($a$) of 0.0345 cm; the kinematic viscosity of water is 0.01 cm$^2$ s$^{-1}$ at 25°C.

**Fig. 8.7** The variation of $I_{\lim}$ with flow rate for the reduction of 1 mM fluorescein (in 0.5 M KCl/0.1 M NaOH) at a mercury wall-jet electrode at (a) pH = 13 ($\times$), and (b) pH = 9.65 ($\circ$). Reproduced from R.G. Compton *et al.*, *J. App. Electrochem.* **20** (1990) 586, with permission from Springer.

Use the pH 13 data to calculate the diffusion coefficient of fluorescein in aqueous solution. Also shown in Fig. 8.7 are data obtained at pH 9.65 using a carbonate/bicarbonate buffer. Suggest, qualitatively, a reason for the altered behaviour and increased currents, as compared to the data at pH 13.

## Solution

Using the expression for $I_{lim}$ in the solution to Problem 8.10 and the respective values for $F$, $v$, $a$, $R$ and [F] ($=10^{-6}$ mol cm$^{-3}$) together with the measured gradient of Fig. 8.7, a value of

$$D = 5.6 \times 10^{-6} \, cm^2 \, s^{-1}$$

is obtained.

The data shown for a pH of 9.65 is seen to follow the simple one-electron behaviour at very fast flow rates, but tends towards two-electron behaviour at low flow rates. Thus the effective number of electrons transferred, $n_{eff}$, changes in a manner expected of an ECE or DISP type process where the one-electron product (S$^{\cdot}$) is rapidly removed from the electrode at fast flows. At low flow rate it is able to react further, leading to the second electron transfer.

It is thought that a DISP 1 process operates in this system:

$$F + e^- \rightleftharpoons S^{\cdot -}$$

$$S^{\cdot -} + H^+ \xrightarrow{slow} SH^{\cdot}$$

$$SH^{\cdot} + S^{\cdot} \rightarrow F + LH$$

where

LH=

Note that the protonation of S$^{\cdot}$ will be faster at a higher proton concentration (lower pH), thus explaining the change in behaviour between pH 13 and pH 9.65.

## 8.12 Wall-Jet Electrode and EC Processes

## Problem

Consider an EC process at a wall-jet electrode:

$$A + e^- \rightleftharpoons B$$

$$B \xrightarrow{k} products$$

The convection–diffusion equations relevant to the EC mechanism in the wall-jet electrode geometry under steady-state conditions are as follows:

$$v_r \frac{\partial [A]}{\partial r} + v_z \frac{\partial [A]}{\partial z} = D \frac{\partial^2 [A]}{\partial z^2} \tag{8.6}$$

$$v_r \frac{\partial [B]}{\partial r} + v_z \frac{\partial [B]}{\partial z} = D \frac{\partial^2 [B]}{\partial z^2} - k[B] \tag{8.7}$$

where $r$ and $z$ are cylindrical coordinates and $D$ is the diffusion coefficient of A and B (assuming these to be equal). The convective solution velocity is described by the two components: $v_r$ for radial velocity, and $v_z$ for the velocity normal to the electrode surface. Close to the electrode, the following approximations can be made [W.J. Albery and C.M.A. Brett, *J. Electroanal. Chem.* **148** (1983) 201]:

$$v_{r,\eta \to 0} \simeq \frac{2}{9} \left( \frac{15M}{2vr^3} \right)^{\frac{1}{2}} \eta$$

$$v_{z,\eta \to 0} \simeq \frac{7}{36} \left( \frac{40Mv}{3r^5} \right)^{\frac{1}{4}} \eta^2$$

where $v$ is the kinematic viscosity and $\eta$ is a dimensionless coordinate:

$$\eta = \left( \frac{135M}{32v^3 r^5} \right)^{\frac{1}{4}} z$$

and $M = k_c^4 V_f^3 / 2\pi^3 a^2$ where $V_f$ is the volume flow rate, $a$ is the nozzle diameter and $k_c$ is a constant determined to be close to 0.90 in many wall-jet systems.

Use the following substitutions

$$\xi = \left( \frac{r}{r_e} \right)^{\frac{9}{8}}$$

$$\chi = \frac{Az}{r^{\frac{7}{8}}}$$

where $A$ is a convenient collection of constants:

$$A = \left( \frac{9C}{8D} \right)^{\frac{1}{3}} r_e^{-\frac{3}{8}}$$

and $r_e$ is the electrode radius and $C = ([5M]^3 / 216v^5)^{-\frac{1}{4}}$ to reduce Eqs. (8.6) and (8.7) to the following dimensionless forms:

$$\frac{\partial^2 [A]}{\partial \chi^2} = \chi \left( \frac{\partial [A]}{\partial \xi} \right)$$

and

$$\frac{\partial^2 [B]}{\partial \chi^2} - K' \xi^{\frac{14}{9}} [B] = \chi \left( \frac{\partial [B]}{\partial \xi} \right)$$

and find an expression for the normalised rate constant $K'$. Comment on the physical significance of the latter expression.

## Solution

First we will substitute $\eta$ into the expressions for $v_{r,\eta \to 0}$ and $v_{z,\eta \to 0}$ and express these quantities in terms of $C$. This is a laborious process of cancelling the rational fractions with the result:

$$v_{r,\eta \to 0} = C \frac{z}{r^{\frac{11}{4}}}$$

$$v_{z,\eta \to 0} = \frac{7}{8} C \frac{z^2}{r^{\frac{15}{4}}}$$

From the definitions of $\xi$ and $\chi$ we can substitute into Eq. (8.6), so long as we take care in determining the correct substitutions for the derivatives:

$$d[A] = \left( \frac{\partial [A]}{\partial \chi} \right)_\xi d\chi + \left( \frac{\partial [A]}{\partial \xi} \right)_\chi d\xi$$

and so

$$\left( \frac{\partial [A]}{\partial z} \right)_r = \left( \frac{\partial [A]}{\partial \chi} \right)_\xi \left( \frac{\partial \chi}{\partial z} \right)_r + \left( \frac{\partial [A]}{\partial \xi} \right)_\chi \left( \frac{\partial \xi}{\partial z} \right)_r$$

$$= \frac{A}{r^{\frac{7}{8}}} \left( \frac{\partial [A]}{\partial \chi} \right)_\xi$$

and

$$\left( \frac{\partial [A]}{\partial r} \right)_z = \left( \frac{\partial [A]}{\partial \chi} \right)_\xi \left( \frac{\partial \chi}{\partial r} \right)_z + \left( \frac{\partial [A]}{\partial \xi} \right)_\chi \left( \frac{\partial \xi}{\partial r} \right)_z$$

$$= \frac{9}{8} \left( \frac{r}{r_e^9} \right)^{\frac{1}{8}} \left( \frac{\partial [A]}{\partial \chi} \right)_\xi - \frac{7}{8} \left( \frac{Az}{r^{\frac{15}{8}}} \right) \left( \frac{\partial [A]}{\partial \chi} \right)_\xi$$

Also

$$\frac{\partial}{\partial z} = \frac{A}{r^{\frac{7}{8}}} \frac{\partial}{\partial \chi}$$

so

$$\frac{\partial^2 [A]}{\partial z^2} = \frac{\partial}{\partial z} \left[ \frac{\partial [A]}{\partial z} \right] = \frac{\partial}{\partial z} \left[ \frac{A}{r^{\frac{7}{8}}} \left( \frac{\partial [A]}{\partial \chi} \right)_\xi \right]$$

$$= \frac{A^2}{r^{\frac{7}{4}}} \left( \frac{\partial^2 [A]}{\partial \chi^2} \right)_\xi$$

Now, substituting for $v_r$, $v_z$, $\frac{\partial[A]}{\partial r}$, $\frac{\partial[A]}{\partial z}$ and $\frac{\partial^2[A]}{\partial z^2}$ into Eq. (8.6) gives

$$
C \frac{z}{r^{\frac{11}{4}}} \frac{9}{8} \left(\frac{r}{r_e^9}\right)^{\frac{1}{8}} \frac{\partial[A]}{\partial \xi} - C \frac{z}{r^{\frac{11}{4}}} \frac{7}{8} \frac{Az}{r^{\frac{15}{8}}} \frac{\partial[A]}{\partial \chi}
$$
$$
+ \frac{7}{8} C \frac{z^2}{r^{\frac{15}{4}}} \frac{A}{r^{\frac{7}{8}}} \frac{\partial[A]}{\partial \chi} = D \frac{A^2}{r^{\frac{7}{4}}} \frac{\partial^2[A]}{\partial \chi^2}
$$

The two terms in $(\partial[A]/\partial \chi)$ cancel, so, multiplying through by $A\, r^{\frac{7}{4}}$:

$$
D A^3 \frac{\partial^2[A]}{\partial \chi^2} = C \frac{Az}{r^{\frac{7}{8}}} \frac{9}{8} \frac{1}{r_e^{\frac{9}{8}}} \frac{\partial[A]}{\partial \xi}
$$

Substituting the expression for $A^3$ and recognising that $\chi$ appears on the right-hand side:

$$
D \frac{9}{8} \frac{C}{D} \frac{1}{r_e^{\frac{9}{8}}} \frac{\partial^2[A]}{\partial \chi^2} = C \chi \frac{9}{8} \frac{1}{r_e^{\frac{9}{8}}} \frac{\partial[A]}{\partial \xi}
$$

$$
\frac{\partial^2[A]}{\partial \chi^2} = \chi \frac{\partial[A]}{\partial \xi}
$$

as required.

Note that overall we have multiplied by $r^{\frac{7}{4}}/(A^2 D)$ to remove dimensionality, after making the substitutions. So for Eq. (8.7)

$$
\frac{\partial^2[B]}{\partial \chi^2} - \frac{kr^{\frac{7}{4}}}{A^2 D} [B] = \chi \frac{\partial[B]}{\partial \xi}
$$

$$
\frac{\partial^2[B]}{\partial \chi^2} - \frac{k}{A^2 D} r_e^{\frac{7}{4}} \xi^{\frac{14}{9}} [B] = \chi \frac{\partial[B]}{\partial \xi}
$$

Hence

$$
K' = \frac{k}{A^2 D} r_e^{\frac{7}{4}}
$$

$$
= \frac{kr_e^{\frac{5}{2}}}{D} \left(\frac{8D}{9C}\right)^{\frac{2}{3}}
$$

$$
= kr_e^{\frac{5}{2}} \left(\frac{64}{81 C^2 D}\right)^{\frac{1}{3}}
$$

The physical significance of $K'$ is that each normalised current/voltage plot in the form of $I/I_{\text{lim}}$ vs $\theta$ $(= F/RT(E - E_f^{\ominus})$ where $E$ is the potential) is a function only of $K'$. In the case of an EC reaction, the effects of the chemically irreversible step after a reversible electron transfer is to shift a reduction to a more positive potential. Thus the shift $F/RT(E_{1/2} - E_f^\ominus)$ between the observed half-wave potential and the formal potential is solely a function of the normalised parameter $K'$.

## 8.13 Sono-Voltammetry

## Problem

A solution of 0.5 mM ferrocene (Fc) in acetonitrile was studied at a platinum electrode of diameter 3 mm in the presence and absence of ultrasound. The ultrasound was provided by a sonic horn positioned opposite to the electrode, which could provide power of up to 60 W cm$^{-2}$.

Under silent conditions, a peak-shaped voltammogram was obtained, whereas under sonication very much larger currents flowed and the voltammogram became sigmoidal and steady state in character, with some spiking seen on the limiting current plateau.

Comment on this, and estimate the approximate diffusion layer thickness under insonation if a limiting current of 314 $\mu$A was observed in the sono-voltammogram. The diffusion coefficient of ferrocene in acetonitrile is $2.3 \times 10^{-5}$ cm$^2$ s$^{-1}$ at 25°C.

## Solution

In quiescent solution, a familiar peak-shaped voltammogram is seen as expected for a macroelectrode under planar diffusion-only transport. In contrast, insonation establishes a strong convective flow as a result of acoustic streaming such that mass transport to the electrode is enhanced significantly. Under the latter conditions, applying the simple Nernst diffusion layer model gives an expression for the limiting current:

$$I_{\text{lim}} = \frac{FAD[\text{Fc}]}{\delta}$$

where $A$ is the electrode area and $\delta$ is the diffusion layer thickness.

Putting $F = 96485$ C mol$^{-1}$, $A = \frac{\pi}{4} \times (0.3)^2$ cm$^2$, $D = 2.3 \times 10^{-5}$ cm$^2$ s$^{-1}$ and $[\text{Fc}] = 0.5 \times 10^{-6}$ mol cm$^{-3}$, we find that for $I = 314$ $\mu$A,

$$\delta = 2.5 \,\mu\text{m}$$

This value is very small compared to the diffusion layer seen at macroelectrodes under silent conditions.

In addition to the strong convection induced by acoustic streaming, if the ultrasound power is above a critical threshold, then cavitation can occur at the electrode–solution interface. This gives rise to the 'spiking' of the observed current as a result of tiny bubble implosions on the electrode.

## 8.14  Rotating Disc Electrodes and Reynolds Number

### Problem

Calculate the Reynolds number for a rotating disc of 5 mm diameter spinning in water and acetonitrile at i) 5 Hz and ii) 25 Hz. Comment on the implication of the values you calculate. The kinematic viscosities of water and acetonitrile at 25°C are $0.01 \text{ cm}^2 \text{ s}^{-1}$ and $4.4 \times 10^{-3} \text{ cm}^2 \text{ s}^{-1}$, respectively.

### Solution

For a rotating disc the Reynolds number is defined as

$$\text{Re} = \frac{W r_e^2}{\nu}$$

where $W$/Hz is the rotation speed, $r_e$ is the disc radius and $\nu$ is the kinematic viscosity.

For water at 5 Hz

$$\text{Re} = \frac{5 \times \left(\frac{0.5}{2}\right)^2}{0.01} \simeq 31$$

whereas at 25 Hz

$$\text{Re} = \frac{25 \times \left(\frac{0.5}{2}\right)^2}{0.01} \simeq 156$$

For acetonitrile at 5 Hz

$$\text{Re} = \frac{5 \times \left(\frac{0.5}{2}\right)^2}{4.4 \times 10^{-3}} \simeq 71$$

and at 25 Hz

$$\text{Re} = \frac{25 \times \left(\frac{0.5}{2}\right)^2}{4.4 \times 10^{-3}} \simeq 355$$

In all cases, Re is well below the threshold for the onset of turbulence. Accordingly, the hydrodynamics of the solution will follow a description assuming *laminar* flow.

## 8.15  Wall-Jet and Rotating Disc Electrodes

### Problem

The one-electron reduction of 3 mM fluoranthene, Fl, to its radical anion was studied in acetonitrile solution containing 0.1 M tetrabutylammonium perchlorate as supporting electrolyte, using a mercury-plated rotating disc electrode of

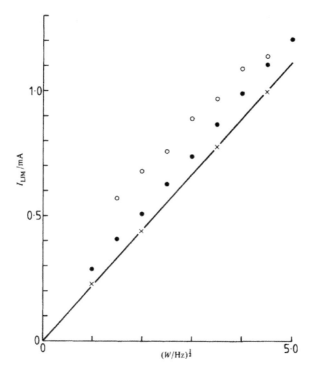

**Fig. 8.8** The variation of the transport-limited current, $I_{\lim}$, with the square root of the disc rotation speed, $W$, for the reduction of fluoranthene. Reproduced from R.G. Compton *et al., J. Chem. Soc., Faraday Trans. 1* **84** (1988) 2013, with permission from the Royal Society of Chemistry.

diameter 0.692 cm. The limiting current was shown to scale linearly with the square root of the disc rotating speed ($W$/Hz) as shown in Fig. 8.8. Calculate the diffusion coefficient of fluoranthene, given that the gradient of the graph in Fig. 8.8 is $2.2 \times 10^{-4}\,\mathrm{A\,s^{\frac{1}{2}}}$.

## Solution

The Levich equation for the transport-limited current, $I_{\lim}$, at a rotating disc electrode is given by

$$I_{\lim} = 1.554 nF\, AD^{\frac{2}{3}} v^{-\frac{1}{6}} [\mathrm{Fl}]_{\mathrm{bulk}} \sqrt{W}$$

where $n$ is the number of electrons transferred, $A$ is the electrode area ($= \pi \times 0.692^2/4$), $D$ is the diffusion coefficient of Fl, $v$ is the kinematic viscosity of acetonitrile ($4.4 \times 10^{-3}\,\mathrm{cm^2\,s^{-1}}$ at 25°C) and $[\mathrm{Fl}]_{\mathrm{bulk}}$ is the bulk concentration of Fl ($3 \times 10^{-6}\,\mathrm{mol\,cm^{-3}}$).

Hence $D$ can be estimated as $1.2 \times 10^{-5}\,\mathrm{cm^2\,s^{-1}}$.

## 8.16 Rotating Disc Electrodes and ECE Processes

## Problem

Consider an ECE process

$$A + e^- \rightleftharpoons B$$

$$B \xrightarrow{k} C$$

$$C + e^- \rightleftharpoons \text{products}$$

taking place at a rotating disc electrode, and suppose both that the A/B process is electrochemically reversible and that species B is so short-lived (unstable) that it is confined to a very thin 'reaction layer' of thickness $\delta_k$ immediately adjacent to the disc surface. In particular, assume $\delta_k \ll \delta_d$ where $\delta_d$ is the diffusion layer thickness.

Hence obtain an expression for $\delta_k$ under steady-state conditions. Also, find an expression for the current–potential waveshape for the ECE process.

## Solution

Under the assumption that the reaction layer is much thinner than the diffusion layer, the transport equation for B can be simplified to

$$\frac{\partial [B]}{\partial t} = D \frac{\partial^2 [B]}{\partial z^2} - k[B]$$

where $D$ is the diffusion coefficient of B and $z$ is the coordinate normal to the electrode. This approximation arises since B does not travel far enough from the electrode to experience convection. This is the 'reaction layer' approximation.

Putting $\partial [B]/\partial t = 0$ for steady state, it follows that

$$[B] = [B]_{z=0} \exp\left(-\frac{z}{\delta_k}\right) \tag{8.8}$$

where

$$\delta_k = \left(\frac{D}{k}\right)^{\frac{1}{2}}$$

which is the first result required.

To obtain the voltammetric waveshape we note first that the electrochemical reversibility of the A/B couple implies

$$[B]_0 = [A]_0 \exp\left(-\theta\right) \tag{8.9}$$

where $[\;]_0$ indicates the concentration at $z = 0$ and

$$\theta = \frac{F}{RT}(E - E^{\ominus}_{f,A/B})$$

and $E^{\ominus}_{f,A/B}$ is the formal potential of the A/B couple; $E$ is the applied electrode potential. Moreover, provided all of the species (A, B, C and products) have the same diffusion coefficient $(D)$, conservation of mass at the electrode surface dictates

$$[A]_0 + [B]_0 + [products]_0 = [A]^* \tag{8.10}$$

where $[A]^*$ is the bulk concentration of A; we have assumed only A to be present in bulk solution. Note that $[C]_0$ is absent from Eq. (8.10) since the electrode reaction of C to products is assumed to be fully driven to the products at the potentials required to reduce A to B.

By Fick's first law the current at a rotating disc electrode is given by

$$\frac{I}{\pi r_e^2 FD} = \frac{\partial[A]}{\partial z}\bigg|_0 - \frac{\partial[products]}{\partial z}\bigg|_0 \tag{8.11}$$

where $r_e$ is the disc radius. Also

$$\frac{\partial[A]}{\partial z}\bigg|_{z=0} = \frac{[A]^* - [A]_0}{\delta_d} \tag{8.12}$$

and

$$\frac{\partial[products]}{\partial z}\bigg|_{z=0} = \frac{[products]_0}{\delta_d} \tag{8.13}$$

where $\delta_d = 0.643 \nu^{\frac{1}{6}} D^{\frac{1}{2}} W^{-\frac{1}{2}}$ and $W/Hz$ is the disc rotation speed and $\nu/cm^2\,s^{-1}$ is the kinematic viscosity of the solution. Further

$$\frac{\partial[A]}{\partial z}\bigg|_{z=0} = -\frac{\partial[B]}{\partial z}\bigg|_{z=0}$$

so that from Eqs. (8.8) and (8.9) we find:

$$\frac{\partial[A]}{\partial z}\bigg|_{z=0} = \frac{[B]_0}{\delta_k} = \frac{[A]_0 \exp(-\theta)}{\delta_k}$$

and hence from Eq. (8.12) we find that

$$[A]_0 = \frac{[A]^*}{1 + \exp(-\theta) \cdot \frac{\delta_d}{\delta_k}} \tag{8.14}$$

From Eqs. (8.9) and (8.10) we have

$$[products]_0 = [A]^* - [A]_0 (1 + \exp(-\theta))$$

and hence from Eqs. (8.11) and (8.14) we find

$$\frac{I}{\pi r_e^2 FD} = \frac{[A]^* - [A]_0 - [products]_0}{\delta_d}$$

$$= \frac{\exp(-\theta)}{\delta_d} [A]_0$$

$$= \frac{[A]^*}{\delta_d} \frac{1}{\exp(\theta) + \frac{\delta_d}{\delta_k}}$$

For very negative potentials

$$\theta \to -\infty, \quad I \to I_{lim}$$

so that

$$\frac{I_{lim}}{\pi r_e^2 FD} = [A]^* \frac{\delta_k}{\delta_d^2}$$

and hence

$$\frac{I_{lim}}{I} = \frac{\delta_k}{\delta_d} \left( \exp(\theta) + \frac{\delta_d}{\delta_k} \right)$$

so

$$\frac{I_{lim}}{I} - 1 = \exp(\theta) \frac{\delta_k}{\delta_d} \tag{8.15}$$

which is the equation describing the current-voltage curve for an ECE process under fast ($k \to \infty$) chemical kinetics, where the A/B step is electrochemically reversible.

It follows from Eq. (8.15) that a mass transport corrected Tafel plot of $E$ versus $\log[I^{-1} - I_{lim}^{-1}]$ will have a gradient of $2.303RT/F$ V per decade.

Finally we emphasise the approximate nature of Eq. (8.15) and stress that in modern work, numerical simulation is used to determine the behaviour of ECE and other reaction mechanisms at rotating disc electrodes without recourse to the reaction layer approximation.

# 9

---

# Voltammetry for Electroanalysis

## 9.1 Electrochemical Sizing of Gold Surfaces

### Problem

Figure 9.1 shows a typical TEM image of a gold nanoparticle modified carbon nanotube. These modified nanotubes may be suspended in solvent. Around 20 $\mu$L of this suspension was deposited onto an electrode and the solvent allowed to evaporate. This method of modifying electrodes is known as 'casting'.

Assume that for the nanoparticles the number of gold atoms per cm$^2$ is the same as that found for a macroscopic polycrystalline gold surface ($1.25 \times 10^{15}$ atoms cm$^{-2}$). Describe a simple method by which the surface area of gold nanoparticles present on an electrode may be measured voltammetrically. Critically assess your suggestion.

### Solution

A cyclic voltammetric experiment run from $+0.0$ to $+1.6$ V (vs SCE) in 0.1 M $H_2SO_4$ allows the oxidation of gold to be clearly resolved, with the gold oxide reduction peak occurring at $+0.85$ V (vs SCE). We assume, as stated in the question, that there are $1.25 \times 10^{15}$ atoms per cm$^2$ on the surface of the gold nanoparticles. The oxidation of polycrystalline gold has been suggested to follow multiple pathways, but overall the process may be viewed as involving a two-electron transfer. The integral of the reduction peak gives a charge which is proportional to the total surface area of gold present. The conversion ratio is $\approx 390$ $\mu$C cm$^{-2}$ [H. Angerstein-Kolowska et al., Electrochimica Acta **31** (1986) 1061]. Figure 9.2 depicts a voltammogram which clearly shows the large gold oxide reduction peak.

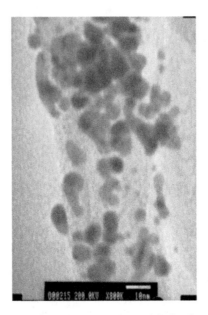

**Fig. 9.1** A typical TEM image of a Au nanoparticle modified carbon nanotube. Reproduced from L. Xiao *et al.*, *Anal. Chim. Acta* **620** (2008) 44, with permission from Elsevier.

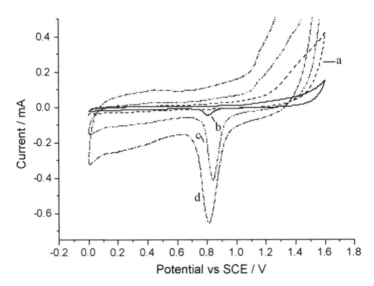

**Fig. 9.2** Cyclic voltammetry of AuCNTs-GC in 0.1 M $H_2SO_4$ with scan rate of $100\,\mathrm{mVs^{-1}}$ with different AuCNTs loadings: (a) blank GC electrode, (b) electrode modified with $20\,\mu\mathrm{L}$ of $0.02\,\mathrm{mg\,mL^{-1}}$ of AuCNTs solution, (c) $20\,\mu\mathrm{L}$ of $0.2\,\mathrm{mg\,mL^{-1}}$ and (d) $40\,\mu\mathrm{L}$ of $0.2\,\mathrm{mg\,mL^{-1}}$. Reproduced from L. Xiao *et al.*, *Anal. Chim. Acta* **620** (2008) 44, with permission from Elsevier.

This analysis, although easy to perform, only provides an estimate of the actual surface area present. This is primarily due to the fact that the surface density of gold atoms on the nanoparticles is likely to be significantly different from that found for macroscopic polycrystalline gold. It has also been implicitly assumed that all the gold nanoparticles are electroactive.

## 9.2  Differential Pulse Voltammetry

### Problem

Linear sweep voltammetric methods often result in poor limits of detection. Pulse voltammetry is frequently used to improve these (see C.M.A. Brett and A.M. Oliveira Brett, *Electroanalysis* (1998) Oxford University Press).

(i) Sketch the potential waveform which is applied to the working electrode during a differential pulse voltammetry (DPV) experiment and give typical values for the pulse amplitudes and times.

(ii) At what points on the potential waveform is the current sampled, and how is the current plotted?

(iii) Explain how DPV allows lower limits of detection to be attained. Your answer should include information on the time scale for the decay of both the Faradaic and capacitive currents.

### Solution

(i) Figure 9.3 shows a schematic diagram of the potential waveform applied to a working electrode during a DPV experiment. The potential amplitude of the pulse ($\Delta E_{pulse}$) is typically of the order of 50 mV and the potential step ($\Delta E_{step}$) is generally 10 mV or less. The timescale of the pulse ($t_{pulse}$) is $\approx$5–100 ms and the duration of the whole step ($t_{step}$) is $\approx$ 0.5–5 s.

(ii) The current is sampled at points 1 and 2, as indicated in Fig. 9.3. From this the voltammogram is plotted as $\Delta I / A \ (= I_2 - I_1)$ vs potential.

(iii) DPV greatly reduces the contribution of non-Faradaic current to the measured voltammetric response. This is possible because capacitive currents in general relax far more quickly than Faradaic processes, specifically within one or two milliseconds. This difference in lifetimes of the two currents may be easily exemplified by considering the capacitive charging of a RC (series resistor-capacitor) circuit as a model of the electrode–solution interface, where $I_{cap} \propto e^{-\frac{t}{RC}}$. This is in contrast to the time dependence of a Faradaic reaction involving a solution phase species (as given by the Cottrell equation for a macroelectrode) where $I_f \propto t^{-\frac{1}{2}}$.

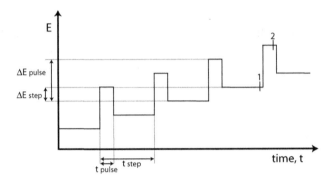

**Fig. 9.3** Schematic waveform for the potential applied to a working electrode during a DPV experiment.

## 9.3 Square-Wave Voltammetry

### Problem

(i) Sketch the potential waveform which is applied to the working electrode during a square-wave voltammetry (SWV) experiment. Give typical values for the pulse amplitudes and times. Indicate on the waveform the points at which the current is sampled. How is an SWV voltammogram plotted?

(ii) How does SWV differ from DPV and why does it in general achieve lower limits of detection?

### Solution

(i) Figure 9.4 shows a schematic diagram of the potential waveform applied to a working electrode during an SWV experiment. $\Delta E_{SW}$ is the amplitude of the pulse ($\approx 25\,\text{mV}$), $\Delta E_{step}$ is the staircase height ($\approx 10\,\text{mV}$) and $t$ is the cycle period ($\approx 5\,\text{ms}$). It is common for the cycle period to be reported as a frequency where $f = 1/t$; consequently, the 'effective' scan rate is then $v_{eff} = f\Delta E_{step}$. The voltammogram is sampled twice each cycle, at points 1 and 2 as marked on Fig. 9.4. The voltammogram is plotted as $\Delta I/A\ (= I_1 - I_2)$ vs potential.

(ii) SWV is a large amplitude pulse technique. Due to this large amplitude it is possible that the reverse pulse can reoxidise the product of the forward pulse (or re-reduce depending upon the scan direction). Consequently, the enhanced sensitivity of SWV results because the net current ($\Delta I$) is larger than either of the currents measured at point 1 or 2.

As with DPV, one of the major advantages of SWV is its ability to cancel non-Faradaic currents as a result of the capacitance being effectively constant at both points 1 and 2. A further advantage of SWV over other pulse techniques

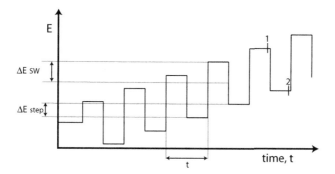

**Fig. 9.4** Schematic waveform for the potential applied to a working electrode during a SWV experiment.

is the far faster scan rates that are possible; this is due to the effectively constant capacitance at both points 1 and 2.

## 9.4 Square-Wave Voltammetry and Dissolved Oxygen

### Problem

Through the use of SWV, how is it possible to selectively minimise some Faradaic currents? In particular, how is it possible to reduce the size of an oxygen signal from an SWV voltammogram recorded in aqueous solution, without degassing?

### Solution

As described in Problem 9.3 part (ii) the increased sensitivity of SWV arises due to the possibility that the analyte of interest can be reoxidised (or re-reduced) on the reverse pulse. Consequently, not only is there an increase in net current, but the reoxidised analyte may then undergo further reduction on the next pulse. This is only possible for analytes which exhibit chemically and electrochemically reversible electron transfer. Consequently, SWV is more sensitive to such species.

The reduction of oxygen in aqueous solutions is irreversible in nature in terms of both the chemical and electrochemical processes involved, and hence it is not reoxidised on the reverse pulse, such that the voltammetric wave as measured is not enhanced. Thus in a solution containing a reversible species of interest and oxygen (where both react electrochemically at similar potentials), the use of SWV may allow analytically useful voltammetric data to be obtained without degassing of the solution. As a caveat, it should be noted that although it is possible to measure an analytically useful voltammetric response in the presence of oxygen, the chemistry occurring within the solution may be altered by its presence.

## 9.5  Stripping Voltammetry

## Problem

(i) Describe how a stripping voltammetry experiment works. Why is it possible to achieve low detection limits ($\approx 10^{-10}$ M)?

(ii) How is it possible to differentiate between different metals contained within one solution?

(iii) What are the advantages and disadvantages of using a mercury-based electrode?

## Solution

(i) A stripping analysis consists of two basic stages. There is first an accumulation stage, and second a voltammetric stripping step induced by sweeping the potential anodically.

For a simple anodic stripping experiment the potential is held at a negative potential so that the metal is reduced onto the working electrode

$$M^{n+}_{(aq)} + ne^- \rightarrow M_{(s)}$$

After a suitable accumulation time the potential is swept in a positive direction so as to oxidise all of the different accumulated metals. The voltage required for the stripping depends on the standard potential for the $M/M^{n+}$ couple and the corresponding electrode kinetics. The magnitude of the stripping peak is then a function of both the concentration of metal ions contained within the solution and the length of the accumulation time. Low detection limits can then be achieved by extending the period of accumulation.

(ii) If a solution contains a mixture of two or more metal ion species, and assuming that they are both deposited onto the electrode during the accumulation stage, then as the electrode is swept positively the metals will be oxidised at their respective potentials. Hence it is possible to identify different metals from the potential at which they are stripped (oxidised) in the voltammetric scan.

Problems may arise when two metals oxidise at similar potentials, so that their voltammetric peaks overlap. If a solid electrode is used, then alloy formation may also present problems. Various methodologies have been developed to overcome this problem: a prime example is the use of mercury film electrodes as opposed to using a hanging mercury drop electrode.

(iii) A huge amount of literature has focused upon the study of stripping analysis, with the majority of work employing a mercury electrode. Mercury electrodes display a number of advantages, including their extended cathodic range in aqueous solution and the fact that upon deposition, the metal may dissolve reversibly into the mercury to form an amalgam, hence allowing far higher levels of accumulation than at a solid electrode.

A major disadvantage is the limited anodic range available, due to the relatively facile oxidation of mercury. A further issue is the known toxicity of mercury and its compounds. In some countries there are stringent controls on its use. As a result of this, there is increasing interest in finding alternative electrode materials. One such example is the use of bismuth (see, for example, J. Wang, *Electroanalysis* **17** (2005) 1341).

## 9.6 Analysis of DNA

### Problem

Adsorptive stripping differs from either cathodic or anodic stripping voltammetry in that the accumulation step for the former does not involve a Faradaic process. Adsorptive stripping voltammetry has helped to increase the range of possible analytes available for detection, including a variety of biomolecules. Pioneering work in the 1960s by E. Paleček developed the use of adsorptive stripping voltammetry for detecting nucleic acids [*Nature* **188** (1960) 656].

 (i) How are the nucleic acids voltammetrically detected? How would the procedure differ if you were using a carbon-based electrode as opposed to a mercury electrode?
(ii) What is: a) DNA hybridisation; and b) a single nucleotide polymorphism?
(iii) Mercury electrodes are highly sensitive to DNA (limits of detection below $0.1\,\mu g\,mL^{-1}$). Why are they of limited use for the detection of DNA hybridisation, and how has this issue been overcome?

### Solution

 (i) Nucleic acids may be voltammetrically detected either through the oxidation or the reduction of the bases. A mercury electrode in aqueous solution has an extended cathodic voltammetric window, but the oxidation of mercury has its onset around $+0.1\,V$ vs SCE. Consequently, the oxidation of the nucleic acids occurs at far too positive potentials for it to be measured at a Hg electrode.

In contrast, a graphite electrode has an electrochemical window between about $-1\,V$ and $+1\,V$ (vs SCE) and as such it is more suited for the measurement of the oxidation of the DNA bases.

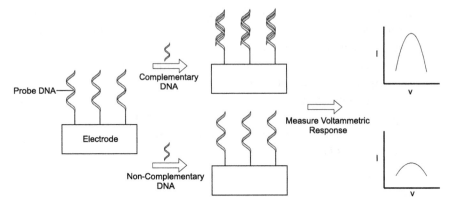

**Fig. 9.5** Schematic diagram of a common methodology for the detection of DNA hybridisation.

(ii) (a) DNA hybridisation is the binding of two complementary single strands of DNA. Detection of the hybridisation event is the aim of most electrochemical DNA sensors.

    (b) A single nucleotide polymorphism is a common form of genome variation where there is a single alteration to the base pair sequence. This single variation is enough to cause an individual to have an increased risk of disease.

(iii) The detection of DNA hybridisation is commonly achieved by modifying an electrode surface with single-stranded DNA. The electrode is then exposed to either a complementary or non-complementary strand of DNA. The hybridisation event may be detected by the increase in the voltammetric signal upon hybridisation. This scheme is outlined in Fig. 9.5. For information on other methodologies of DNA detection, see the work of C. Batchelor-McAuley and R.G. Compton [*Biosensors and Bioelectronics* **24** (2009) 3183].

DNA has a high affinity for mercury. As such the single-stranded DNA may bind non-specifically to the electrode, leading to false positive results. This problem may be circumvented through the use of magnetic beads such that the hybridisation event occurs on the surface of the beads. The beads may be easily transferred to another solution where the hybrid DNA is dissociated from the magnetic beads. The concentration of DNA in this resulting solution may then be measured using adsorptive stripping voltammetry at a Hg electrode. For more information on the use of magnetic beads for biosensing see the work of E. Paleček and M. Fojta [*Talanta* **74** (2007) 276].

## 9.7 The Clark Cell

## Problem

(i) Describe the Clark sensor used for the amperometric sensing of gaseous oxygen.

(ii) Show that the steady-state limiting flux (mol cm$^{-2}$ s$^{-1}$) to the cathode of a Clark cell is given by

$$j_{lim} = \frac{p_{O_2}}{\left(\frac{d_e}{P_e}\right) + \left(\frac{d_m}{P_m}\right)} \tag{9.1}$$

where $P_m$ and $P_e$ are the oxygen permeabilities in the membrane and electrolyte layer, respectively, and $p_{O_2}$ is the prevailing partial pressure of oxygen in the sample. $d_e$ and $d_m$ are the thicknesses of the electrolyte and membrane layers, respectively. Note that the permeability, $P$, is defined as $P \propto \alpha D$, where $\alpha$ is the oxygen solubility and $D$ is the oxygen diffusion coefficient.

(iii) In a typical practical example, $d_m = 20\ \mu m$, $d_e = 5\ \mu m$, $P_m = 8 \times 10^{-11}$ mol $m^{-1}$ s$^{-1}$ atm$^{-1}$ and $P_e = 2.7 \times 10^{-9}$ mol $m^{-1}$ s$^{-1}$ atm$^{-1}$ [C.E.W. Hahn, *Analyst* **123** (1998) 57R]. How can Eq. 9.1 be simplified? Comment on the implications for the practical use of Clark sensors.

## Solution

(i) The basis of the Clark sensor is shown in Fig. 9.6. The sensor is built around a working electrode which brings about the reduction of oxygen either to water or to hydrogen peroxide, depending on the nature of the electrode material. In the case of a Pt cathode, water will be formed. An Ag/AgCl electrode acts as a reference electrode and as the counter electrode. The electrolyte is typically an aqueous buffer solution to which Cl$^-$ ions have been added; a pH near 7 is generally chosen. A gas-permeable membrane separates the electrolyte from the oxygen sample.

(ii) Consider a coordinate, $x$, normal to the electrode surface where $x = 0$ is the cathode, $x = d_e$ is the electrolyte/membrane boundary and $x = d_e + d_m$ is the membrane/sample boundary.

Since the cathode is held at a potential corresponding to the diffusion-controlled reduction of oxygen, the concentration of oxygen at $x = 0$ is $c = 0$. Let $c_2'$ be the oxygen concentration in the electrolyte at $x = d_e$, $c_2$ be the corresponding value at $x = d_e$ in the membrane, and $c_3$ the concentration at $x = d_e + d_m$ inside the membrane. Accordingly, when

$$x = d_m + d_e, \quad c_3 = \alpha_m p_{O_2}$$

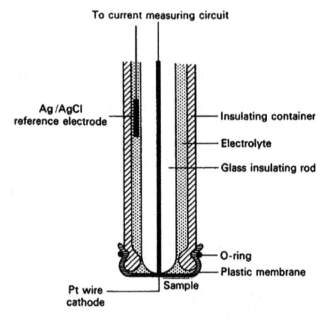

**Fig. 9.6** Schematic outline of the Clark $p_{O_2}$ sensor. Reproduced from C.E.W. Hahn, *Analyst* **123** (1998) 57R, with permission from the Royal Society of Chemistry.

and

$$x = d_e, \quad \frac{c_2'}{c_2} = \frac{\alpha_e}{\alpha_m}$$

The flux, $j$, is given by

$$j_{\lim} = \frac{D_m}{d_m}(c_3 - c_2) \quad \text{(membrane)}$$

$$= \frac{D_e}{d_e}(c_2') \quad \text{(electrolyte)}$$

where the equality is required by conservation of mass across the membrane–electrolyte boundary.

Solving the above equations, and eliminating $c_3$ and $c_2'$, we see

$$c_2 = \frac{c_3}{1 + \frac{D_e \alpha_e}{D_m \alpha_m} \frac{d_m}{d_e}}$$

$$= \frac{c_3}{1 + \frac{P_e d_m}{P_m d_e}}$$

Hence

$$j_{\lim} = \frac{D_m}{d_m}(c_3 - c_2)$$

$$= \alpha_m p_{O_2} \frac{D_m}{d_e} \left( \frac{1}{1 + \frac{P_m d_e}{P_e d_m}} \right)$$

$$= \frac{p_{O_2}}{\left( \frac{d_e}{P_e} \right) + \left( \frac{d_m}{P_m} \right)}$$

as required.

(iii) Considering the parameters given, it is clear that $(d_m/P_m) \gg (d_e/P_e)$ in a typical situation. Hence, the equation for the flux simplifies to

$$j_{lim} \simeq \frac{P_m}{d_m} \cdot p_{O_2}$$

indicating that transport through the membrane, rather than the electrolyte, is the slower step.

Practically speaking, this has the undesirable consequence that the sensor response is highly sensitive to the membrane thickness $(d_m)$ and membrane properties $(P_m)$. Calibration of the sensor is thus essential.

## 9.8 Calibration and Limits of Detection

## Problem

(i) Describe a general methodology by which a voltammetric sensor can be used to determine the concentration of an analyte *without* prior calibration.

(ii) Define the term 'limit of detection'.

(iii) The anodic stripping voltammetry of consecutive additions of Pb(II) $(2 \, \mu g \, L^{-1})$ to a degassed 0.1 M acetate buffer solution was recorded at an edge-plane pyrolytic graphite electrode. From the following data, find the limit of detection of Pb(II) for this system.

| Concentration/$\mu g \, L^{-1}$ | Peak Area/$\mu C$ |
|:---:|:---:|
| 2.0 | 1.24 |
| 4.0 | 3.19 |
| 6.0 | 4.96 |
| 8.0 | 6.57 |
| 10.0 | 8.67 |
| 12.0 | 10.20 |
| 14.0 | 12.20 |
| 16.0 | 14.20 |
| 18.0 | 16.20 |
| 20.0 | 19.00 |

## Solution

(i) Assuming there is a linear response between the concentration of the analyte and the voltammetric signal, a sample of unknown concentration may be readily analysed through the use of the standard addition method.

First, the voltammetric response of a solution of unknown concentration is measured. This solution is then repeatedly spiked with a known concentration of the analyte and the voltammetric response recorded after each addition. From the intercept of a plot of the response versus the additional concentration of the solution, the initial solution concentration can be found. An example of such a plot is shown in Fig. 9.7, where the intercept (as indicated on the graph) shows that the initial analyte concentration was 3 mM.

(ii) The limit of detection is the minimum concentration of an analyte which results in a significantly different response from that of a blank signal. There are a number of ways of calculating this value, but a common procedure is to calculate the standard deviation ($y$-residuals) of a calibration plot and the limit of detection ($3\sigma$) is taken as being equal to three times the standard deviation divided by the gradient of the plot.

(iii) Figure 9.8 shows the required plot of concentration versus peak area. From this it is possible to find the line of best fit (as shown). The standard deviation of the $y$-residuals from this line of best fit is found to be 0.34 $\mu$C and the

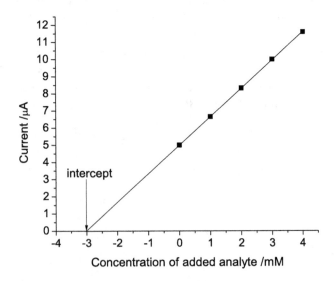

Fig. 9.7 Example of a standard addition plot for the determination of an unknown concentration of analyte.

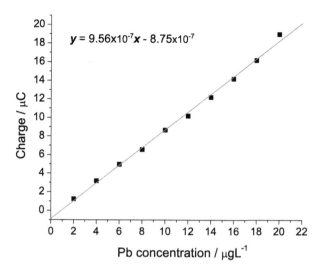

**Fig. 9.8** Calibration plot for the anodic stripping voltammetry of Pb(II) at an edge-plane pyrolytic graphite electrode. Reproduced from M. Lu *et al.*, *Electroanalysis* **23** (2011) 1089, with permission from Wiley.

gradient of the line of best fit is $0.96 \, \mathrm{C \, g^{-1} \, L}$.

$$\text{Limit of detection} = \frac{3\sigma}{\text{gradient}}$$
$$= \frac{3 \times 0.36}{0.96}$$

Hence the limit of detection of Pb(II) is found to be $1.1 \, \mu\mathrm{g \, L^{-1}}$.

## 9.9 Enzyme Electrodes

### Problem

Enzyme electrodes are the focus of a huge amount of research, partially due to their innate selectivity and sensitivity towards a specific analyte. An enzyme catalysed reaction may be simply described as below

$$\mathrm{E + S} \underset{k_{-1}}{\overset{k_1}{\rightleftharpoons}} \mathrm{ES} \overset{k_2}{\rightarrow} \mathrm{E + P} \tag{9.2}$$

where E is the enzyme, S is the substrate, P is the product and ES is an intermediate complex. Derive an equation which describes how the rate of reaction varies with the concentration of the analyte S. Comment on what implications your result has for using enzyme electrodes for analytical purposes.

## Solution

The derivation in question is the commonly used Michaelis–Menten equation. This requires us to assume that the total concentration of the intermediate ES is at steady state such that

$$\frac{d[ES]}{dt} = k_1[E][S] - k_{-1}[ES] - k_2[ES] = 0$$

The rate of formation of the product is given by

$$\frac{d[P]}{dt} = k_2[ES]$$

In order to solve this we must additionally appreciate that the total concentration of the enzyme is constant, so

$$[E]_{tot} = [ES] + [E]$$

This allows us to find an expression for the concentration of $[ES]$

$$[ES] = \frac{[S][E]_{tot}}{K_M + [S]}$$

where

$$K_M = \frac{k_{-1} + k_2}{k_1}$$

Consequently,

$$\frac{d[P]}{dt} = k_2[ES] = k_2 \frac{[S][E]_{tot}}{K_M + [S]}$$

so

$$\frac{d[P]}{dt} = \frac{V_m[S]_{tot}}{K_M + [S]} \tag{9.3}$$

where

$$V_m = k_2[E]_{tot}$$

$K_M$ is the Michaelis–Menten constant, and $V_m$ is the maximum rate of the reaction. Equation 9.3 shows us that there is only a limited range of substrate concentrations for which the enzyme electrode will be analytically useful (as shown in Fig. 9.9).

## 9.10  Glucose Biosensors

## Problem

One of the major successes of electroanalytical chemistry is the development and production of glucose biosensors. These devices have had a huge impact upon

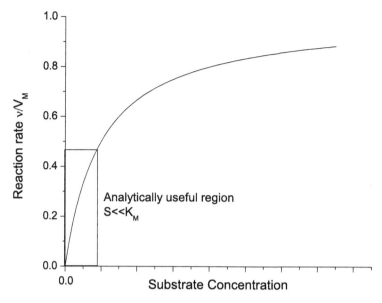

**Fig. 9.9** Dependence of the enzyme catalysed reaction rate with substrate concentration [S].

the everyday management of diabetes for those able to afford the purchase of the sensors. A recent review on the subject can be found in the work of J. Wang [*Chem. Rev.* **108** (2008) 814].

(i) Describe how the original *first-generation* glucose sensors operate.

(ii) Why are these sensors sensitive to the oxygen concentration?

(iii) Discuss how the so-called *second-generation* glucose sensors avoid issues associated with oxygen.

## Solution

(i) The detection of glucose is based upon the following reaction:

$$\text{glucose} + O_2 \xrightarrow{\text{glucose oxidase}} \text{gluconic acid} + H_2O_2$$

The original glucose sensor was based upon measuring the concentration of oxygen consumed by the above reaction. This measurement is achieved through the use of a Clark cell (see Problem 9.7). This methodology was further improved by electrochemically monitoring the production of hydrogen peroxide via its oxidation at a platinum electrode.

(ii) Due to the above-described sensors being reliant on the indirect sensing of glucose via detection of oxygen, they are also susceptible to errors arising from fluctuations in the oxygen concentration. This problem is further compounded by the fact that the oxygen concentration in a sample can be far

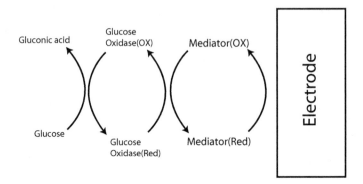

**Fig. 9.10** Schematic diagram showing the redox processes involved in a second-generation glucose sensor.

lower than that of glucose. Hence the sensors may become oxygen deficient, and as such an incorrectly low level of glucose concentration is measured.

(iii) A number of approaches have been proposed in order to help avoid the issues associated with using oxygen as a mediator. The most prominent is the use of non-physiological electron acceptors such as ferrocene or its derivatives. These electron acceptors shuttle the electrons between the glucose under analysis and the electrode, as shown schematically in Fig. 9.10. It should be noted that it is not possible for glucose oxidase to directly transfer electrons to the electrode, due to the redox centre being buried within the enzyme.

## 9.11  Detection of Vitamin B$_{12}$

### Problem

Figure 9.11 (trace A) shows voltammetry for vitamin B$_{12}$ in aqueous buffer at a basal-plane graphite electrode. The wave corresponds to the two-electron reduction of the Co(III) centre in vitamin B$_{12}$ to Co(I):

$$Co(III)L_{(aq)} + 2e^- \rightarrow Co(I)L_{(aq)}$$

The electrode was then modified by depositing tiny droplets of water-insoluble 1,2-dibromocyclohexane (DBCH) on the electrode. The altered voltammetry is shown in Fig. 9.11 (trace C); in the absence of vitamin B$_{12}$ in solution no voltammetric response from DBCH is seen (trace B).

(i) What is a basal-plane graphite electrode? Comment on why this substrate might have been selected for the experiment.

(ii) Explain the voltammetric response in trace C. Comment on the sites of electron transfer in the scheme you propose.

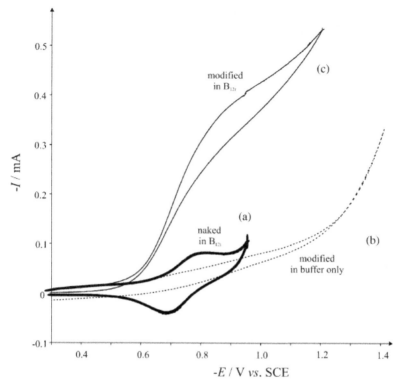

**Fig. 9.11** Basal-plane pyrolytic graphite electrode in vitamin $B_{12r}$ at a scan rate of $0.5\,V\,s^{-1}$ for (a) unmodified BPPG in 1.2 mM vitamin $B_{12r}$ solution, (b) modified BPPG with DBCH in buffer solution, and (c) modified BPPG in vitamin $B_{12r}$ solution. Reproduced from P. Tomčik *et al.*, *Anal. Chem.* **76** (2004) 161, with permission from the American Chemical Society.

(iii) How might the experiment be adapted to provide an analytical method for the quantification of vitamin $B_{12}$ in pharmaceutical products and biological matrix media?

## Solution

(i) Basal-plane pyrolytic graphite (BPPG) electrodes and edge-plane pyrolytic graphite (EPPG) electrodes are fabricated from highly ordered pyrolytic graphite (HOPG), as shown in Fig. 6.1. The latter consists of layers of graphite with an interlayer spacing of 3.35 Å. If the graphite crystal is cut parallel to the graphite sheets, a BPPG electrode results, whereas cutting in a perpendicular direction gives an EPPG electrode. Of course, although the BPPG surface is flat relative to an EPPG surface it is not perfectly flat; there are steps on the surface separating terraces of basal-plane graphite.

The choice of a BPPG electrode for the experiment described is partly since the material needs to be inactive with respect to the reduction of DBCH. The reduction of allylbromides has a high overpotential on carbon electrodes as compared to, say, silver electrodes. Additionally, the flat surface allows droplets of the oily DBCH to be formed on the electrode surface.

(ii) Trace C shows that the reduction peak in the voltammetry of Co(III)L is greatly increased in the presence of the modifying DBCH, but that the reverse peak (re-oxidation of Co(I)L) is markedly diminished. This is typical of electrocatalytic behaviour (EC′) and suggests that the following net chemical step is involved:

$$2Co(I)L_{(aq)} + RBr_{2(oil)} \xrightarrow{2H^+} 2Co(II)L_{(aq)} + 2Br^-_{(aq)} + R'_{(oil)}$$

where $RBr_2$ is DBCH, R′ is cyclohexane and Co(II)L is Vitamin $B_{12}$. At the potential of the forward voltammetric wave, Co(II)L will be reduced to Co(I)L, so completing the catalytic cycle.

Considering the sites of electron transfer, the reactions

$$Co(III)L_{(aq)} + e^- \rightarrow Co(II)L$$

and

$$Co(II)L + e^- \rightarrow Co(I)L$$

must occur on the remaining exposed basal plane terraces (not covered by the DBCH droplets), whilst the catalytic reaction must take place on the surface of the droplets, with the Co(II)L species diffusing back to the basal plane to complete the catalytic cycle.

(iii) Analytically, a carbon paste electrode can be made with the DBCH acting as the paste binder and a reactive material facilitating the electrocatalytic detection of the target. The paste is formed by mixing carbon powder with DBCH and empirically optimising the relative amounts for optimal analytical performance [P. Tomčik *et al.*, *Anal. Chem.* **76** (2004) 161].

## 9.12 The Anodic Stripping Voltammetry of Industrial Effluent

## Problem

The detection of metals in industrial effluents is important for environmental monitoring. One approach for the detection of copper(II) ions is through the use of anodic stripping voltammetry (ASV). Figure 9.12 shows ASV voltammograms performed in an effluent of pH 2.1, with a chemical oxygen demand (COD) of

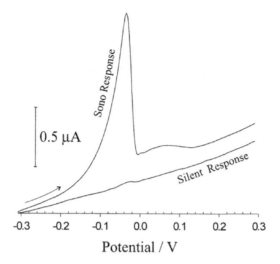

**Fig. 9.12** Linear sweep voltammograms detailing the copper stripping responses at a bare glassy carbon electrode after ultrasonically enhanced deposition and after conventional silent deposition. Deposition time: 30 s, scan rate: $50\,mV\,s^{-1}$. Reproduced from J. Davis, *Analytical Letters* **34** (2001) 2375, with permission from Taylor & Francis.

1963 mg $L^{-1}$ and containing 22.6 ppm pesticide, 6.7 ppm herbicide, trace acetone, methanol and xylene. Copper was pre-concentrated on a glassy carbon electrode at a voltage of $-1.0\,V$ (vs SCE) after the addition of concentrated KCl to the effluent. After a period of 30 seconds of deposition, the potential was scanned in a positive direction to produce the trace shown in Fig. 9.12. The responses shown are those seen under silent conditions and under insonation.

 (i) Explain the principles of ASV as applied in this experiment.
 (ii) Suggest a reason for the addition of excess chloride ions.
(iii) Why are much larger signals seen under insonation than under silent conditions?
(iv) How might signals such as these in Fig. 9.12 be used to give analytically useful data?

## Solution

 (i) ASV is a two-step procedure. In the first step, the Cu(II) analyte is preconcentrated as Cu(0) on the electrode by reduction at $-1.0\,V$:

$$\frac{1}{2}Cu^{2+}_{(aq)} + e^- \rightleftharpoons \frac{1}{2}Cu_{(s)}$$

After 30 seconds, when a significant deposit of $Cu_{(s)}$ has built up on the glassy carbon electrode, the potential is then scanned towards positive values, so bringing about the reoxidation of the copper in the second step of ASV.

The size of this 'stripping' signal is used to quantify the amount of material deposited on the electrode, and hence the amount of Cu(II) in solution.

(ii) Excess chloride ion ($Cl^-$) is added, as KCl, to provide a controlled chloride concentration environment since Cu(II) and Cu(I) both form complexes with $Cl^-$ ions, e.g. $CuCl_4^{2-}$, $CuCl_2^-$, etc. Accordingly, the stripping signal can change both in size and potential under varying concentrations of chloride. By adding a large excess of KCl, reproducible stripping signals are seen.

(iii) Under silent conditions, components in the effluent adsorb on the electrode surface, thus preventing the deposition of the metallic copper. Most probably it is the organic material in the effluent which is responsible. Insonation leads to the *in situ* cleaning of the electrode: cavitational collapse at the surface temporarily removes adsorbate so allowing the deposition of copper. In addition, the very strong convection promoted by the insonation greatly enhances the rate at which Cu(II) is brought to the electrode surface for deposition. This improves the sensitivity of the ASV technique by greatly increasing the extent of pre-concentration.

(iv) In order to extract analytically useful data, standard additions — known small volumes of fixed Cu(II) concentration — would be added to the sample and the increase in the peak current or area (C) would be plotted against the added Cu(II) concentration (in the sample) as shown in Fig. 9.13. The intercept on the $x$-axis corresponds to the unknown Cu(II) concentration in the sample.

## 9.13   Adsorptive Stripping Voltammetry at Carbon Nanotube Modified Electrodes

### Problem

Explain the technique of adsorptive stripping voltammetry (AdSV). Carbon electrodes are often modified with a layer of carbon nanotubes to increase the sensitivity of AdSV. Suggest a reason why, and discuss the types of analytical targets that might most usefully benefit from this approach.

### Solution

In the AdSV technique, the analytical target is first pre-concentrated on the electrode surface by allowing the electrode to stand in the analytical solution, either

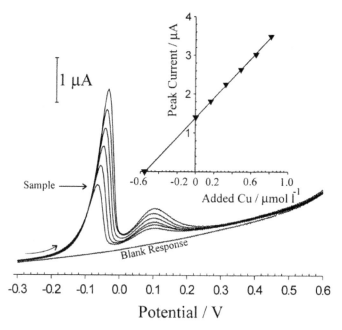

Fig. 9.13 'SonoASV' determination of copper within sample 1. Deposition time: 30 s, scan rate: 50 mV s$^{-1}$. Inset: corresponding standard addition calibration. Reproduced from J. Davis, *Analytical Letters* **34** (2001) 2375, with permission from Taylor & Francis.

at open circuit or at a well-defined potential, for a period of tens of seconds or minutes. This allows the target to adsorb onto the surface. After a period of time, usually experimentally optimised, the electrode potential is swept in either an oxidising or reducing direction to obtain a Faradaic signal from the adsorbed analyte. The signal size (peak current and/or peak area) is then used to infer the quantity of material in the original sample, either by the method of standard additions or by using electrodes with well-defined adsorption properties so that a single scan can be used to generate a quantitative signal without the need for standard addition.

Multi-walled carbon nanotubes (MWCNTs) are electronically conductive and also show a very large surface area per unit length. Accordingly, casting these onto an electrode surface greatly increases the surface area for adsorption whilst retaining electrical contact between the nanotube surface and the underlying electrode. As a result the sensitivity (current per unit concentration) is significantly and usefully increased. Electrodes made of screen-printed carbon nanotubes can be used similarly.

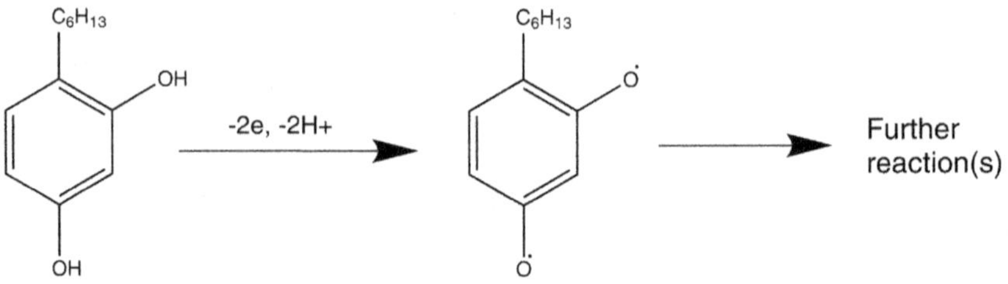

**Fig. 9.14** Mechanism for the electrochemical oxidation of 4-hexylresorcinol. Reproduced from R.T. Kachoosangi *et al.*, *Electroanalysis* **20** (2008) 1714, with permission from Wiley.

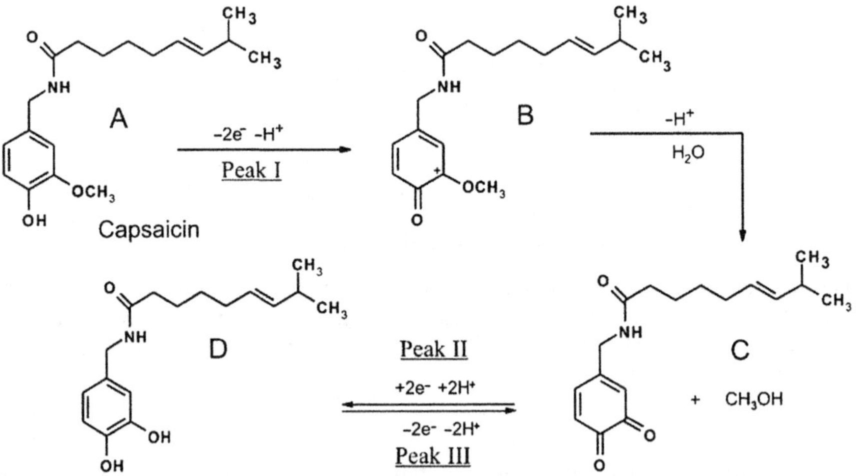

**Fig. 9.15** Mechanism for the electrochemical oxidation/reduction of capsaicin. Reproduced from R.T. Kachoosangi *et al.*, *Analyst* **133** (2008) 888, with permission from the Royal Society of Chemistry.

Adsorption onto the MWCNTs is favoured strongly by $\pi - \pi$ interactions between the aromatic surface of the CNTs and any aromatic moieties in the target molecule. Two examples of the use of MWCNTs for adsorption stripping voltammetry are as follows:

(i) 4-hexylresorcinol in pharmaceutical products [R.T. Kachoosangi *et al.*, *Electroanalysis* **20** (2008) 1714] where the target is oxidised via the mechanism shown in Fig. 9.14 and the associated Faradaic current detected.

(ii) The detection of capsaicin and hence the 'heat' of chilli peppers. Figure 9.15 shows the mechanism and the electrochemistry underlying the Faradaic response [R.T. Kachoosangi *et al.*, *Analyst* **133** (2008) 888].

## 9.14 Surface Modified Electrodes

## Problem

Electrodes are often surface modified in an attempt to improve their 'electroanalytical responses'. Comment critically on the implications of modifying a macroelectrode in each of the following ways:

(i) Modification with a monolayer of adsorbed material which mediates electron transfer to or from an analyte.
(ii) Modification with a partial coverage of nanoparticles.
(iii) Modification with a porous layer of either $C_{60}$ or carbon nanotubes.

## Solution

(i) If the electrode is modified with a monolayer, or less, of adsorbed molecules, then the transport to the electrode will be essentially unmodified and, in the case of a macroelectrode, will be described by simple semi-infinite planar diffusion. The effect of the monolayer mediating the electron transfer will be to change the overall electrode kinetics of the oxidation or reduction of the analytical target at the interface, ideally with the effect of accelerating the process.

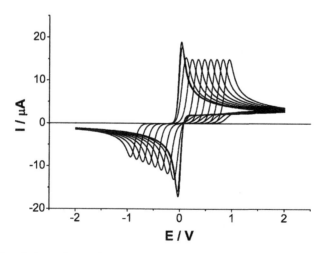

**Fig. 9.16** Simulation of one-electron oxidation process. $\alpha = \beta = 0.5$, $D_A = D_B = 10^{-5} \, cm^2 \, s^{-1}$, $E_f^{\ominus} = 0 \, V$, $v = 0.1 \, V \, s^{-1}$, $A = 0.0707 \, cm^2$, $C_A^* = 1 \, mM$. $k_0$ ranges from 1 cm $s^{-1}$ to $10^{-10}$ cm $s^{-1}$. Reproduced from B.R. Kozub *et al.*, *Sens. and Actuators B* **143** (2010) 539, with permission from Elsevier.

That said, the effect on the size of the analytical signal will not be significant if linear sweep voltammetry (LSV) is used to identify the presence of the target. Figure 9.16 shows the effect of changing electrode kinetics from very slow to very fast corresponding to the electrochemically fully irreversible to the fully reversible limit. The peak current changes, but not significantly. In fact the ratio of the peak currents for a one-electron reduction is given by the ratio of the corresponding Randles–Ševčík equations for LSV at a macroelectrode:

$$\frac{I_{p,rev}}{I_{p,irrev}} = \frac{2.69 \times 10^5 Ac^* D^{\frac{1}{2}} v^{\frac{1}{2}}}{2.99 \times 10^5 \alpha^{\frac{1}{2}} Ac^* D^{\frac{1}{2}} v^{\frac{1}{2}}}$$
$$= \frac{2.69}{2.99\alpha^{\frac{1}{2}}}$$

where $A$ is the electrode area, $c^*$ is the analyte concentration, $D$ is the diffusion coefficient and $v$ is the voltammetric scan rate. For the case of $\alpha = 0.5$

$$\frac{I_{p,rev}}{I_{p,irrev}} = \frac{2.69}{2.99\alpha^{\frac{1}{2}}} = 1.27$$

so that the measured effect of accelerating the electrode kinetics is a maximum of 30%, although, of course, the peak potential moves to lower overpotentials which may be analytically useful in removing the effects of some interfering species. Note, however, that if pulse techniques are used rather than LSV, then the acceleration of electrode kinetics from irreversible to reversible behaviour can lead to a greater improvement in analytical signal.

(ii) The effect of partially covering an electrode surface with a random array of nanoparticles depends on the degree of coverage of the surface. At very low coverages, and assuming that electrolysis occurs only on the surface of the nanoparticles and not at the underlying electrode surface, then diffusion occurs to essentially independent nanoparticles so that a sigmoidal voltammogram with a constant limiting current is observed at low scan rates, turning into a peak-shaped response at faster scan rates. The signal size (current) reflects the number of nanoparticles present.

As the coverage increases, there is increasing overlap of the diffusion fields (Fig. 9.17) of the nanoparticles and eventually a signal is observed that is very close to that expected if the entire geometric area of the electrode *were active*. Note that not all of the electrode needs to be covered by nanoparticles for this to occur.

The different possible diffusional cases are summarised in Fig. 9.17 and are discussed in Chapters 6 and 11 in more detail. In Case 4 there is effectively linear diffusion to the entire geometric area covered by the nanoparticles, and so the peak-shaped voltammetry reflects the area of the underlying electrode

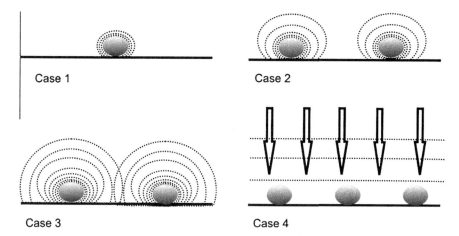

**Fig. 9.17** Diffusion at a nanoparticle array. Case 1: almost planar diffusion at an isolated nanoparticle where the diffusion layer thickness is small compared to the nanoparticle radius. Case 2: convergent diffusion to diffusionally independent nanoparticles. Case 3: partially overlapping diffusion layers between adjacent nanoparticles. Case 4: heavily overlapping diffusion layers leading to effectively linear diffusion to the array as a whole. Reproduced from Y.G. Zhou *et al.*, *Chem. Phys. Lett.* **497** (2010) 200, with permission from Elsevier.

supporting the nanoparticles. Accordingly, the maximum current that can be obtained by linear diffusion to this area is achievable with only partial coverage of the electrode surface with nanoparticles. Hence the method is very useful for economising on the cost of expensive metals such as platinum, gold or palladium.

The above discussion assumes implicitly that the nanoparticles have similar electrochemical behaviour to the corresponding bulk metal. In fact, the changing surface and electronic structures between the bulk material and nanoparticle means that there may be qualitatively altered electroreactivity; at least in principle some reactions may not proceed at the nanoscale, or vice versa (see also Chapter 11).

(iii) Covering the electrode with a porous layer of $C_{60}$ or carbon nanotubes causes a change in mass transport regime from semi-infinite linear diffusion to quasi 'thin-layer' voltammetry. Larger currents are often seen in the linear sweep voltammetry from this effect alone, as explained in greater detail in Problem 11.6. In principle, the $C_{60}$ or carbon nanotubes may mediate an electron transfer at a different rate from that due to direct transfer from the underlying electrode, but this is difficult to prove (or disprove) on the basis of voltammetry using layers of $C_{60}$ or nanotubes, because of the significant mass transport effects described above. This is discussed in more detail in Problem 11.6.

## 9.15 Electron Transfer Rates at Carbon Electrodes

### Problem

The extent of the electrochemical reversibility of analyte signals at carbon electrodes varies from one type of electrode to another: edge-plane pyrolytic graphite (EPPG), basal-plane pyrolytic graphite (BPPG), glassy carbon (GC) and boron-doped diamond (BDD) all show different responses. What are the analytical benefits of varying (and optimising) the electrode materials for a particular task, assuming detection via linear sweep voltammetry?

### Solution

By altering the rate of electron transfer, the signals for the analytes can be made to occur closer to the formal potential for the reaction of interest, as the electrode kinetics become faster. If the voltammetric signal changes from electrochemically irreversible to reversible there will be a small increase in the peak current (see Problem 9.14) for linear sweep voltammetry but the most important consequence is the decreased overpotential at which current can be driven. Hence, less extreme potentials are required for the analysis.

Examples include the reduction of chlorine

$$\frac{1}{2}Cl_{2(aq)} + e^- \rightleftharpoons Cl^-_{(aq)}$$

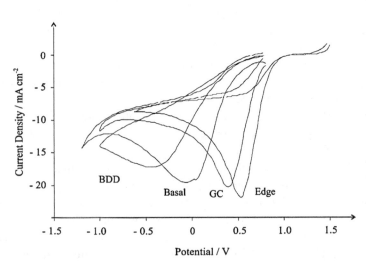

**Fig. 9.18** Cyclic voltammograms for the reduction of chlorine in 0.1 M nitric acid solution using EPPG (0.196 cm$^2$), GC (0.07 cm$^2$), BPPG (0.196 cm$^2$), and BDD (0.07 cm$^2$) electrodes. All scans were recorded at 100 mV s$^{-1}$, vs. SCE. Reproduced from E.R. Lowe *et al.*, *Anal. Bioanal. Chem.* **382** (2005) 1169, with permission from Springer.

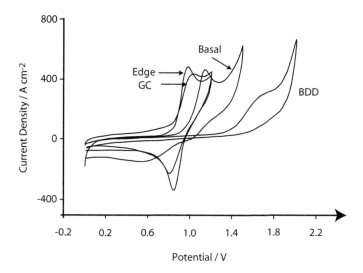

**Fig. 9.19** The electrochemical oxidation of bromide (1 mM) at EPPG (dotted line), BPPG, GC and BDD electrodes, in a 0.1 M nitric acid solution. All scans recorded at 100 mV s$^{-1}$, vs. SCE. Reproduced from E.R. Lowe *et al.*, *Electroanalysis* **17** (2005) 1627, with permission from Wiley.

as shown in Fig. 9.18 and the oxidation of bromide

$$Br^-_{(aq)} \rightleftharpoons \frac{1}{2} Br_{2(aq)} + e^-$$

as shown in Fig. 9.19. In each case it can be seen that EPPG shows the fastest electrode kinetics and hence least overpotential for the reactions of interest.

In general, optimisation of electrode materials for analytical tasks is desirable and worthwhile, while noting that *sensitivity* in terms of peak current (but not peak potential) in linear sweep voltammetry is essentially controlled by the rate of diffusion to the electrode and hence, provided a signal is seen at some potential, relatively insensitive to the electrode material.

# 10

---

# Voltammetry in Weakly Supported Media: Migration and Other Effects

## 10.1 Coulomb's Law

### Problem

Coulomb's law states that the electrostatic force, $|\mathbf{F}|$, acting on two charges $q_1$ and $q_2$ separated by a distance $r$ is given by:

$$|\mathbf{F}| = \frac{1}{4\pi\epsilon_s\epsilon_0} \frac{q_1 q_2}{r^2}$$

(i) How does this law indicate that opposite charges attract and like charges repel?
(ii) Define $\epsilon_0$ and $\epsilon_s$.
(iii) In water at 298 K, the constant of proportionality is:

$$\frac{1}{4\pi\epsilon_s\epsilon_0} = \frac{1}{4\pi \times 78.54 \times 8.854 \times 10^{-12}} = 1.14 \times 10^8 \, \mathrm{J\,m\,C^{-2}}$$

Hence calculate the attractive force on a proton, $H^+$, due to a chloride ion, $Cl^-$, at a displacement of 1 nm.
(iv) The potential energy on the proton associated with this attractive force is given by integration of Coulomb's law as:

$$U = \frac{1}{4\pi\epsilon_s\epsilon_0} \frac{q_1 q_2}{r}$$

Calculate the energy *per mole* associated with this interaction. Assess its magnitude.

## Solution

(i) If the charges are opposite, the sign of the force will be negative, and therefore the force will act to accelerate the charges in the direction of decreasing $r$, according to Newton's second law. Since $r$ indicates the separation of the charges, decreasing $r$ brings the charges towards each other, and hence they attract. If both charges have the same sign, the force acts in the direction of increasing $r$, which is repulsion.

(ii) $\epsilon_0$ is the permittivity of free space. It indicates the strength of electrostatic interactions as a proportion of the interacting charges.

$\epsilon_s$ is the relative permittivity of the medium as a proportion of $\epsilon_0$, sometimes called the 'dielectric constant' of the medium. It indicates the extent to which electrostatic interactions are screened by the inherent polarity of a medium, as compared to vacuum. In water, which is a highly polar solvent, it has the value 78.54 at 298 K, indicating that electrostatic attractions are approximately eighty times *weaker* in water due to screening by the solvent than is the case in vacuum.

(iii) Substituting into the equation, and denoting the charge on an electron as $e$:

$$|F| = 1.14 \times 10^8 \times \frac{1 \times -1 \times e^2}{(10^{-9})^2}$$
$$= -1.14 \times 10^{26} \times e^2$$
$$= 2.94 \times 10^{-12}\,\mathrm{N}$$

(iv) The energy is $r$ times the force, so over 1 nm distance this is $2.94 \times 10^{-21}$ J. Multiplying by the Avogadro constant to determine the energy per mole, we find

$$U \times N_A = 2.94 \times 10^{-21} \times 6.022 \times 10^{23} = 1.77\,\mathrm{kJ\,mol^{-1}}$$

Note that at 298 K, $RT$ is approximately 2.4 kJ mol$^{-1}$. Therefore, Coulombic attractions have considerable magnitude over nanoscale distances. This explains why an understanding of electrodynamics is particularly vital to the study of electrochemistry at the nanoscale (see also Chapter 11).

## 10.2 The Nernst–Planck Equation

### Problem

Ignoring the effects of non-ideality, the electrochemical potential, $\mu_i$, of an ion in solution may be defined:

$$\mu_i = \mu_i^0 + RT \ln c_i + z_i F \phi$$

where $\mu_i^0$ is the electrochemical potential at a chosen standard state, $R$ is the gas constant, $T$ is temperature, $c_i$ is concentration, $z_i$ is the charge of the ion, $F$ is the Faraday constant and $\phi$ is potential.

(i) Given that

$$|F| = -\frac{d\mu_i}{dx}$$

determine the force, $|F|$, acting on a mole of ions as a function of the local concentration and potential.

(ii) If the frictional force is proportional to velocity with a constant of proportionality $f_i$, derive an expression for the limiting velocity $v_i$. Show that the flux corresponding to this velocity is given by the Nernst–Planck equation (Eq. 10.1), for a linear direction.

$$J = -D_i \left( \frac{dc_i}{dx} + \frac{z_i F}{RT} c_i \frac{d\phi}{dx} \right) \qquad (10.1)$$

(iii) By defining the mobility, $u_i$, as the ratio of the limiting speed to the applied electric field (with no concentration gradients), determine the relationship between $u_i$ and the diffusion coefficient $D_i$.

(iv) Determine an expression for $\partial c_i / \partial t$, given that in a linear space conservation of mass requires:

$$\frac{\partial c_i}{\partial t} = -\frac{\partial J_i}{\partial x}$$

## Solution

(i) From the equation given, we see that force is the negative first derivative of energy with respect to a space coordinate. This is because 'force' describes a tendency to travel down an energy gradient to a point of minimum energy. Since the chemical potential is defined as the Gibbs energy per mole, the force per mole is the first derivative of the chemical potential:

$$|F| = -\frac{d\mu_i}{dx} = -RT \frac{d\ln c_i}{dx} - z_i F \frac{d\phi}{dx}$$

(ii) A limiting velocity means that acceleration is zero, and so overall force is zero according to Newton's second law. Therefore, the frictional force on a single molecule must exactly balance the force due to the chemical potential gradient:

$$\frac{-1}{N_A} \left( RT \frac{d\ln c_i}{dx} + z_i F \frac{d\phi}{dx} \right) - f_i v_i = 0$$

therefore, rearranging

$$v_i = \frac{-1}{N_A f_i} \left( RT \frac{d \ln c_i}{dx} + z_i F \frac{d\phi}{dx} \right)$$

The flux $J$ through a plane is the product of the velocity of the ions undergoing transport, and the concentration of ions:

$$J = v_i c_i$$

$$= - \left( \frac{RT}{N_A f_i} c_i \frac{d \ln c_i}{dx} + \frac{z_i F}{N_A f_i} c_i \frac{d\phi}{dx} \right)$$

$$= - \left( \frac{RT}{N_A f_i} \frac{dc_i}{dx} + \frac{z_i F}{N_A f_i} c_i \frac{d\phi}{dx} \right)$$

which is recognisably similar to the Nernst–Planck equation. If we express the diffusion coefficient as the ratio

$$D_i = \frac{RT}{N_A f_i}$$

then we can derive the more common form:

$$J = -D_i \left( \frac{dc_i}{dx} + \frac{z_i F}{RT} c_i \frac{d\phi}{dx} \right)$$

as was given in Eq. 10.1.

(iii) The electric field, E, is $-d\phi/dx$. Therefore if the concentration gradient is zero:

$$u_i = -\frac{v_i}{-\frac{d\phi}{dx}}$$

$$= \frac{z_i F}{N_A f_i}$$

Typically, $|z_i|$ is used such that absolute speed rather than velocity (implying a direction) is considered. Hence from the definition of $D_i$ we can confirm that:

$$u_i = D_i \frac{|z_i| F}{RT}$$

which is often referred to as the Einstein relation.

(iv) Care must be taken in differentiating the expression for $J$ since both $\phi$ and $c_i$ are dependent on the space coordinate and so the product rule must be used. The correct derivative is:

$$\frac{\partial c_i}{\partial t} = -\frac{\partial J}{\partial x} = D_i \left( \frac{\partial^2 c_i}{\partial x^2} + \frac{z_i F}{RT} \left( c_i \frac{\partial^2 \phi}{\partial x^2} + \frac{\partial c_i}{\partial x} \frac{\partial \phi}{\partial x} \right) \right) \tag{10.2}$$

Clearly this equation is cumbersome by comparison to Fick's second law, which is the special case where $\partial \phi / \partial x = 0$ everywhere.

Most importantly, the Nernst–Planck equation is non-linear, whereas Fick's second law is linear. Both analytical theory and simulation using the full Nernst–Planck equation are hence much more demanding than diffusion-only theory, thus indicating one good reason why excess supporting electrolyte is often used to remove the influence of electric fields in a system.

## 10.3 Migration and the Electric Field

## Problem

In a generalised geometry, the Nernst–Planck equation is given for a species $i$ as

$$J_i = -D_i \left( \nabla c_i + \frac{z_i F}{RT} c_i \nabla \phi \right)$$

where $\nabla$ indicates the gradient operator and $J_i$ is a flux vector.

On the right-hand side, the first term describes diffusion and the second term describes migration.

(i) What is migration?

(ii) The electric field is defined as:

$$E = -\nabla \phi$$

By considering the Nernst–Planck equation, discuss the meaning of the concept of 'electric field'.

(iii) Gauss's law states that

$$\nabla \cdot E = -\frac{\rho}{\epsilon_s \epsilon_0}$$

where $\nabla \cdot$ is the divergence operator, $\rho$ is the charge density, and $\epsilon_s \epsilon_0$ is the absolute permittivity of the medium.

Use this equation to relate $\phi$ to the ion concentrations.

(iv) Why is the relation derived here necessary, in addition to the Nernst–Planck equation (Eq. 10.2), in order to describe migrational processes?

(v) Why is Coulombic attraction or repulsion not observed in fully supported solutions, even though there is a much higher concentration of ions than in weakly supported solutions?

## Solution

(i) Migration is the *organised* movement of an ionic species in response to Coulombic forces, i.e. by attraction to or repulsion from other ions or a charged surface. Since opposite charges attract, migration will tend to compensate the development of charge separation in a solution. Because the cause for migration is Coulombic attraction, it can be seen as a tendency towards local equilibrium by minimisation of energy, by maximising ionic attraction. By contrast, diffusion is a random process by which the system tends to local equilibrium by maximisation of entropy.

The contributions of diffusion and migration are described respectively by the two terms in the right-hand side of the Nernst–Planck equation derived above (Eq. 10.1).

(ii) An electric field (given as a vector E) is a mathematical formulation of the average Coulombic force acting on an ion at a point, due to the spatial distribution of other charges. The direction of the electric field is given such that a positive ion will travel down the field, while a negative ion will travel up the field. As well as arising from ion distribution, applying a charge to an electrode surface will cause an electric field in the vicinity, since ions will be attracted to that charge.

(iii) The electric field due to a charge distribution is described mathematically by Gauss's law, one of Maxwell's equations:

$$\nabla \cdot \mathbf{E} = -\frac{\rho}{\epsilon_s \epsilon_0}$$

This law is equivalent to Coulomb's law above, but is written in terms of a charge density rather than in terms of the locations of individual ions, so it is much easier to work with when the number of ions is large.

In an ionic solution, the ions are typically the only source of charge density, and so this charge density is the sum of the charge density due to each ion. This charge density is the product of the ionic charge, which is a multiple $z_i$ of the electron charge $e$, and the ionic number densities, denoted $n_i$:

$$\rho = e \sum_i z_i n_i$$

$$= F \sum_i z_i c_i$$

where $F$ is the Faraday constant ($=N_A e$) and $c_i$ is a concentration. Since the electric field can be described as the negative gradient of a potential, $\phi$, such that $\mathbf{E} = -\nabla \phi$, we can write:

$$\nabla^2 \phi = -\frac{F}{\epsilon_s \epsilon_0} \sum_i z_i c_i$$

which is commonly called the Poisson equation. This describes the potential field arising from a certain charge distribution.

(iv) One Nernst–Planck equation applies for every species in the solution. If we are to solve the equations for $n$ species, these contain $n + 1$ unknowns, which are the $n$ concentrations $c_i$ and the potential $\phi$. An additional relation is therefore required to solve this set of equations. This relation is often chosen to be the Poisson equation, since this is derived from Maxwell's equations and therefore relates potential and concentration in a manner that is self-consistent and which agrees with Coulomb's law.

Historically, another common choice for a further constraint has been the electroneutrality approximation. This uses the small absolute value of $\epsilon_0$ to approximate that:

$$\sum_i z_i c_i = 0$$

Note that this approximation is not accurate at short timescales or across short distances, where 'short' means nanoscale, in practice.

(v) Note that migration refers to organised motion due to Coulombic forces. An individual ion will instantaneously be subject to Coulombic forces in one direction, as ions move with respect to each other, but these forces are averaged to zero over a very short timescale. Especially if the concentration of ions in solution is large, the repulsive and attractive forces in different directions will cancel each other out. Therefore, there is no lasting force acting in one direction, so on average an ion is unaffected by these forces. Migration is therefore not observed when the ionic strength is high, unless an electric field has been applied to the solution.

## 10.4 Transport Numbers and Liquid Junction Potentials

## Problem

(i) Explain the term transport number as applied to an electrolyte $M^{z+}X^{z-}$, and explain why a liquid junction potential arises at the boundary between two liquids containing different concentrations of MX, if the cation and anion have different transport numbers.

(ii) Show how the following concentration cell *with transport* can be used to find the transport number of $Cu^{2+}_{(aq)}$ ion:

$$Pb_{(s)} | PbSO_{4(s)} | CuSO_{4(aq, \, 0.2 \, M)} | CuSO_{4(aq, \, 0.02 \, M)} | PbSO_{4(s)} | Pb_{(s)}$$

for which the cell emf is +0.0118 V. Assume that the activity coefficients of $SO_4^{2-}$ in 0.2 M and 0.02 M solution are 0.110 and 0.320, respectively [J.D.R. Griffiths and P.J.F. Thomas, *Calculations in Advanced Physical Chemistry* (1971) Arnold].

## Solution

(i) An electrochemical current in bulk solution is carried by both the cation ($M^{z+}$) and anion ($X^{z-}$). These generally move at different speeds under the influence of an electric field, so that it is useful to introduce the concept of transport numbers, $t_+$ and $t_-$, which are the fraction of current carried by the cation and by the anion, respectively. In terms of the molar conductivities, $\Lambda_+$ and $\Lambda_-$

$$t_+ = \frac{\Lambda_+}{\Lambda_+ + \Lambda_-}$$

$$t_- = \frac{\Lambda_-}{\Lambda_+ + \Lambda_-}$$

where by definition $t_+ + t_- = 1$.

A liquid junction potential arises at the interface because in the case that $t_+ \neq t_-$, the different rates of diffusion of the two ions from high to low concentration generates a charge separation at the interface and hence a potential difference known as a liquid junction potential, $E_{LJP}$. According to the classical theory of liquid junction potentials [P. Henderson, *Z. Physik. Chem.* **59** (1907) 118], for a salt $M^{z+}$ and $X^{z-}$, this has a limiting magnitude of

$$E_{LJP} = (t_+ - t_-)\frac{RT}{zF}\ln\frac{a_{LHE}}{a_{RHE}}$$

where $a_i$ is the activity on the side $i$ of the junction, and the potential difference is measured as the difference between the right- and left-hand electrodes (RHE and LHE).

(ii) In the case of the concentration cell with transport, the measured potential can be viewed as the sum of the reversible concentration cell with potential, $E_{rcc}$, and the liquid junction potential. The former can be established by considering the processes at the two electrodes:

RHE

$$\frac{1}{2}PbSO_{4(s)} + e^- \rightleftharpoons \frac{1}{2}SO^{2-}_{4(aq,\ 0.02\ M)} + \frac{1}{2}Pb_{(s)}$$

LHE

$$\frac{1}{2}PbSO_{4(s)} + e^- \rightleftharpoons \frac{1}{2}SO^{2-}_{4(aq,\ 0.2\ M)} + \frac{1}{2}Pb_{(s)}$$

Net

$$\frac{1}{2}SO^{2-}_{4(aq,\ 0.2\ M)} \rightleftharpoons \frac{1}{2}SO^{2-}_{4(aq,\ 0.02\ M)}$$

So that

$$E_{\text{rcc}} = -\frac{RT}{F} \ln \frac{a^{1/2}_{SO_4^{2-}, \text{RHE}}}{a^{1/2}_{SO_4^{2-}, \text{LHE}}}$$

$$= -\frac{RT}{2F} \ln \frac{a_{SO_4^{2-}, \text{RHE}}}{a_{SO_4^{2-}, \text{LHE}}}$$

and

$$E_{\text{LJP}} = -(t_+ - t_-)\frac{RT}{2F} \ln \frac{a_{SO_4^{2-}, \text{RHE}}}{a_{SO_4^{2-}, \text{LHE}}}$$

$$= (1 - 2t_+)\frac{RT}{2F} \ln \frac{a_{SO_4^{2-}, \text{RHE}}}{a_{SO_4^{2-}, \text{LHE}}}$$

where we have used the relation that $t_+ + t_- = 1$.
Hence

$$E = E_{\text{rcc}} + E_{\text{LJP}}$$

$$= (-1 + 1 - 2t_+)\frac{RT}{2F} \ln \frac{a_{SO_4^{2-}, \text{RHE}}}{a_{SO_4^{2-}, \text{LHE}}}$$

$$= t_+ \frac{RT}{F} \ln \frac{a_{SO_4^{2-}, \text{LHE}}}{a_{SO_4^{2-}, \text{RHE}}}$$

$$= t_+ \frac{RT}{F} \ln \frac{0.20 \times 0.110}{0.02 \times 0.320}$$

$$= 0.0118 \, \text{V}$$

It follows that

$$t_+ = \frac{0.0118F}{RT} \frac{1}{\ln \frac{0.20 \times 0.110}{0.02 \times 0.320}}$$

$$= \frac{0.4951}{1.235}$$

$$= 0.372$$

and so $t_+ \approx 0.37$ (and $t_- \approx 0.63$).

## 10.5 Transport Numbers and the Hittorf Method

## Problem

A solution of lithium chloride was electrolysed for a long period in a cell of two components – a cathode compartment and an anode compartment – separated with a glass frit so that there was transport between the two solutions across the interface defined by the frit (a 'cell with transport'). Each compartment contained a platinum electrode; at the anode, chloride anions were oxidised, whereas at the cathode, water was reduced.

(i) Explain how analysis of the chloride content of the anode compartment before and after electrolysis can be used to find the transport number of $Cl^-$ (and hence $Li^+$). This is the classical 'Hittorf' method for finding transport numbers.

(ii) In an experiment the passage of $0.05\ F$ in charge caused the mass of LiCl in the anode compartment to decrease by $0.708\ g$. Estimate the transport numbers of $Li^+$ and $Cl^-$. The molecular mass of LiCl is $42.394\ g\ mol^{-1}$.

## Solution

(i) In the anode compartment, electrolysis oxidises a certain amount of chloride ion corresponding to a charge, $+Q$, passed through the cell. So, $n_{el} = Q/F$ moles of chloride are lost at the electrode. Equivalently, the reduction of water in the cathode compartment passes a corresponding charge $-Q$.

This transfer is mediated by the passage of an ionic current across the cell. An amount of $Cl^-$ corresponding to this charge $(t_- n_{el})$ moves from the cathode compartment to the anode compartment; at the same time an amount of $Li^+$ $(t_+ n_{el})$ moves from the anode compartment to the cathode compartment, as shown in Fig. 10.1 for the hypothetical case of $t_+ = 0.75$ and $t_- = 0.25$.

In the anode compartment, $t_+ n_{el}$ moles of $Li^+$ are lost; equally $n_{el}$ moles of $Cl^-$ are consumed by electrolysis, but this is replenished by a current carrying $t_- n_{el}$ moles of $Cl^-$, so the overall change in $Cl^-$ molarity is

$$(1 - t_-)n_{el} = -t_+ n_{el}$$

which is the same as the change in $Li^+$ molarity. This makes sense because electroneutrality must be maintained in both compartments.

Hence, after the electrolysis, the loss of LiCl in the anode compartment corresponds to $t_+ n_{el}$, which is the proportion of the charge passed that is carried by the ionic current of $Li^+$. If this loss is denoted $\Delta n_{LiCl}$, then

$$t_+ = \frac{\Delta n_{LiCl}}{n_{el}} = \frac{\Delta n_{LiCl}}{Q/F} \tag{10.3}$$

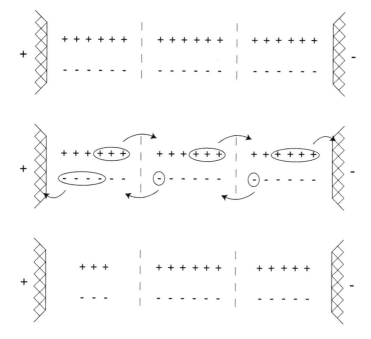

**Fig. 10.1** Measurement of transport numbers via the Hittorf method.

(ii) A mass of 0.708 g corresponds to (0.708/42.934) moles of LiCl. This evaluates as $1.67 \times 10^{-2}$ moles. The charge of $0.05\,F$ corresponds to $5 \times 10^{-2}$ moles which have undergone electrolysis.

Using the expression for $t_+$ at Eq. 10.3

$$t_+ = \frac{1.67 \times 10^{-2}}{5 \times 10^{-2}} = 0.334$$

Hence $t_+ \approx 0.33$, and $t_- \approx 0.67$ for the salt LiCl.

## 10.6 The Gouy–Chapman Equation

## Problem

In *Understanding Voltammetry*, the problem of the diffuse double layer as conceived in the Gouy–Chapman theory is posed as:

$$\frac{\partial^2 \Theta}{\partial \chi^2} = \sinh \Theta \qquad (10.4)$$

where $\Theta$ is a normalised potential and $\chi$ is a normalised length scale.

(i) In order to describe a diffuse double layer at the surface of an electrode charged to a potential $\Theta_0$, what boundary conditions apply? Explain the terms 'potential of zero charge' and 'ideally polarisable electrode'.

(ii) Use the Nernst–Planck equation at steady state to find steady-state concentration profiles of a binary salt $A^+X^-$ in a diffuse double layer with a potential function $\Theta(\chi)$. Explain your result physically.

(iii) Using the identity

$$\frac{\partial^2 \Theta}{\partial \chi^2} = \frac{1}{2} \frac{\partial}{\partial \Theta} \left( \frac{\partial \Theta}{\partial \chi} \right)^2$$

determine the function $\Theta(\chi)$. The following double angle formulae for hyperbolic functions will also be helpful:

$$\sinh(2x) = 2 \sinh x \, \cosh x$$
$$\cosh(2x) = 2 \sinh^2 x + 1$$

(iv) Show that this corresponds to a potential decaying exponentially away from the electrode, when $\Theta_0 \ll 1$.

## Solution

(i) If the surface of the electrode is at $\chi = 0$, then $\Theta = \Theta_0$ at this boundary. In bulk solution, which occurs at $\chi \to \infty$, $\Theta = 0$, since we may choose bulk solution to have a reference potential which is arbitrarily zero. The potential difference between the electrode and bulk is caused by the potentiostat applying a charge to the electrode and hence changing its Fermi level. If the reference electrode is ideal, a change to the potential difference applied between the working and reference electrodes will correspond exactly to the change in potential difference between the working electrode and bulk solution.

The 'potential of zero charge' of the electrode is the potential difference between the working and reference electrodes at which the potential difference between the working electrode and bulk solution is zero, such that the working electrode is uncharged. An 'ideally polarisable electrode' (IPE) is one where altering this charge does not induce any Faradaic current (a reaction) or adsorption, but simply induces migration of ions in solution.

(ii) At steady-state in linear space, conservation of mass requires that:

$$-\frac{\partial J_i}{\partial \chi} = \frac{\partial c_i}{\partial t} = 0$$

Hence, the flux $J_i$ of any species $i$ is constant through space. Since for an ideally polarisable electrode, the flux of any species into or out of the electrode must be zero (since the species does not absorb, adsorb or react at this boundary), $J_i = 0$ everywhere. Therefore from the Nernst–Planck equation

$$\frac{\partial c_i}{\partial \chi} + z_i c_i \frac{\partial \Theta}{\partial \chi} = -J_i = 0$$

If $c_i$ is non-zero in bulk, we can assume that it is not zero anywhere, and therefore it is valid to divide through both sides of the equation by $c_i$:

$$\frac{1}{c_i} \frac{\partial c_i}{\partial \chi} = \frac{\partial \ln c_i}{\partial \chi} = -z_i \frac{\partial \Theta}{\partial \chi}$$

As $\chi \to \infty$, $c_i \to c^*$ and $\Theta \to 0$. Therefore:

$$\int_\chi^\infty \frac{\partial \ln c_i}{\partial \chi'} \, d\chi' = -z_i \int_\chi^\infty \frac{\partial \Theta}{\partial \chi'} \, d\chi'$$

$$[\ln c_i]_{\chi=\chi'}^{\chi=\infty} = -z_i \, [\Theta]_{\chi=\chi'}^{\chi=\infty}$$

$$\ln c^* - \ln c_i = -z_i(-\Theta)$$

$$c_i = c^* \exp{(-z_i\Theta)}$$

This is another way of deriving the Boltzmann distribution.

The elevation of the concentration of negative ions at an electrode of positive potential occurs because the positive charge on the electrode attracts anions and repels cations, thus forming an equilibrated double layer where the net negative charge on this layer exactly cancels the positive charge on the electrode.

The exponential nature of the accumulation of oppositely charged ions means that Gouy–Chapman theory predicts local concentrations which are much larger than that in bulk if the potential difference applied between the electrode and the solution greatly exceeds $RT/F$. In practice, the Stern layer of adsorbed solvent molecules and the specific adsorption of ions mediates the potential perceived by the diffuse component of the double layer to a great extent.

(iii) Substituting the identity given above into the equation to be solved:

$$\frac{1}{2} \frac{\partial}{\partial \Theta} \left( \frac{\partial \Theta}{\partial \chi} \right)^2 = \sinh \Theta$$

Clearly this is integrable. We can identify that since $\Theta \to 0$ as $\chi \to \infty$, $\partial\Theta/\partial\chi \to 0$ in the same limit, and so, marking variables of integration with ':

$$\int_0^{\left(\frac{\partial\Theta}{\partial\chi}\right)^2} d\left(\frac{\partial\Theta}{\partial\chi}\right)'^2 = 2\int_0^{\Theta} \sinh\Theta'\, d\Theta'$$

$$\left(\frac{\partial\Theta}{\partial\chi}\right)^2 = 2\left[\cosh\Theta'\right]_0^{\Theta}$$

$$= 2(\cosh\Theta - 1)$$

From the double angle formula for $\cosh x$, and noting that $\cosh\Theta = \cosh(2(\Theta/2))$:

$$\left(\frac{\partial\Theta}{\partial\chi}\right)^2 = 4\sinh^2\left(\frac{\Theta}{2}\right)$$

$$\frac{\partial\Theta}{\partial\chi} = \pm 2\sinh\left(\frac{\Theta}{2}\right)$$

This is separable and so can be integrated to determine $\Theta$ as a function of $\chi$. We will integrate towards the inner boundary ($\chi = 0$), where $\Theta$ is known:

$$\int_{\Theta_0}^{\Theta} \frac{d\Theta'}{2\sinh\left(\frac{\Theta'}{2}\right)} = \pm\int_0^{\chi} d\chi$$

$$= \pm\chi$$

The integral on the left can be solved by using the double angle formula for $\sinh x$, then multiplying top and bottom by $\cosh(\Theta/4)$ and integrating by an appropriate substitution. This works because the substitutions make it clear that we are integrating the ratio of a function to its antiderivative, a common pattern.

$$\int_{\Theta_0}^{\Theta} \frac{d\Theta'}{2\sinh\left(\frac{\Theta'}{2}\right)} = \int_{\Theta_0}^{\Theta} \frac{d\Theta'}{4\sinh\left(\frac{\Theta'}{4}\right)\cosh\left(\frac{\Theta'}{4}\right)}$$

$$= \int_{\Theta_0}^{\Theta} \frac{\cosh\left(\frac{\Theta'}{4}\right) d\Theta'}{4\sinh\left(\frac{\Theta'}{4}\right)\cosh^2\left(\frac{\Theta'}{4}\right)}$$

$$= \int_{\Theta_0}^{\Theta} \frac{\operatorname{sech}^2\left(\frac{\Theta'}{4}\right) d\Theta'}{4\tanh\left(\frac{\Theta'}{4}\right)}$$

Now if $u = \tanh(\Theta/4)$, $du = \frac{1}{4}\operatorname{sech}^2(\Theta/4)\,d\Theta$:

$$\int_{\tanh\left(\frac{\Theta_0}{4}\right)}^{\tanh\left(\frac{\Theta}{4}\right)} \frac{du}{u} = [\ln u]_{\tanh\left(\frac{\Theta_0}{4}\right)}^{\tanh\left(\frac{\Theta}{4}\right)}$$

$$= \ln\left(\frac{\tanh\left(\frac{\Theta}{4}\right)}{\tanh\left(\frac{\Theta_0}{4}\right)}\right)$$

Therefore,

$$\frac{\tanh\left(\frac{\Theta}{4}\right)}{\tanh\left(\frac{\Theta_0}{4}\right)} = \exp(\pm\chi)$$

The sign is resolved by recognising that as $\chi \to +\infty$, $\Theta \to 0$ and therefore $\tanh(\Theta/4) \to 0$. Therefore, the sign must be negative:

$$\tanh\left(\frac{\Theta}{4}\right) = \tanh\left(\frac{\Theta_0}{4}\right)\exp(-\chi)$$

as required.

(iv) If $\Theta_0 \ll 1$, and recognising that $|\Theta| \leq |\Theta_0|$, it is appropriate to expand $\tanh(\Theta/4)$ using its Taylor series about $\Theta = 0$, which to the first order is:

$$\tanh x \approx x$$

Hence

$$\frac{\Theta}{4} \approx \frac{\Theta_0}{4}\exp(-\chi)$$

$$\Theta \approx \Theta_0\exp(-\chi)$$

which is an exponentially decaying potential outwards from $\chi = 0$.

## 10.7 Ohmic Drop

### Problem

Voltammetry for the oxidation of ferrocene in increasingly weakly supported solutions of acetonitrile is shown at Fig. 10.2. The scan rate is 200 mV s$^{-1}$ and the electrode radius is 0.3 mm. Answer the following about the voltammetry.

(i) Why does the peak-to-peak separation increase with lower support ratio?
(ii) Why does the formal potential for the redox couple of ferrocene and ferrocenium appear to shift to positive potential at lower support ratio?
(iii) Why do we expect no significant change to forward peak currents due to migration at low support?
(iv) Would you expect such a dramatic effect with a microelectrode?

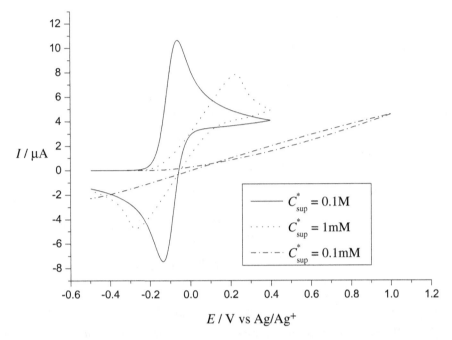

$E$ / V vs Ag/Ag$^+$

**Fig. 10.2** Simulated cyclic voltammetry for the oxidation of 3 mM ferrocene at a 0.3 mm radius Au hemisphere in different concentrations of supporting electrolyte (tetra-*n*-butylammonium perchlorate) in acetonitrile.

## Solution

(i) As the support ratio is reduced, the conductivity of the solution is lowered, and therefore the solution is more resistive. Hence, to drive an equivalent current, a greater overpotential is required. Therefore, the overpotential required to drive sufficient current for mass transport to become rate-limiting is increased, compared to fully supported voltammetry.

This is reflected in the voltammetry by each peak being more separated from the formal potential, and so the overall peak-to-peak separation is larger at lower support.

(ii) Recall that the formal potential contains a term equal to:

$$\frac{RT}{F} \ln \prod_i \gamma_i^{\nu_i}$$

as was shown in Chapter 1 in the derivation of the Nernst equation. The $\gamma_i$ are the activity coefficients of the respective species. For charged species, such as the ferrocenium cation in the ferrocene–ferrocenium couple, $\gamma_i$ is a strong function of the ionic strength due to the interaction of the ion with an ionic atmosphere provided by the other ions in solution.

At low concentrations of supporting electrolyte, the Debye–Hückel limiting law (see also Problem 1.6) is appropriate for determining the magnitude of $\gamma_i$, but at ionic strengths as large as required for fully supported voltammetry, other effects such as electrostriction begin to become important, and it is difficult to predict the variation of $\gamma_i$ with concentration. Nonetheless, an ionic strength dependency is expected, and is observed.

(iii) In a transport-limited current regime, where migration would be significant, the current is controlled by the transport of the electroactive ferrocene species to the electrode surface. Since ferrocene is neutral, it is not subject to Coulombic forces and therefore its transport is unaffected by migration, irrespective of the electric field.

(iv) At a microelectrode, the currents are considerably lower. Therefore, the $IR$ term associated with ohmic drop is much smaller. At low current, the quantity of charge injected into solution is negligible compared to the ease of transport of the supporting species. The voltammetry is then less strongly affected by support ratio, although of course at very low support we still expect microelectrode voltammetry to be distorted due to the presence of electric fields.

## 10.8 The Zero-Field Approximation

### Problem

(i) What is the 'zero-field approximation' in the study of weakly supported voltammetry? Explain why the zero-field approximation is likely to be correct for a microelectrode, and why it is useful.

(ii) Why is the zero-field approximation unlikely to be accurate for a nanoelectrode?

### Solution

(i) We have already seen that at steady state, the double layer accumulates a charge, by migration of ions, that exactly opposes the charge on the electrode itself. Therefore, from a point outside the double layer in a spherically symmetric space, the charge on the electrode is fully screened and no Coulombic forces act on ions outside the double layer in the absence of any further impetus for charge separation.

So, the electric field at the outer boundary of the double layer is zero; this is formalised by Gauss's law (given by Eq. 10.5 in a hemispherically symmetric space) which states that the electric field across a surface is proportional to the charge enclosed within the surface. When the net enclosed charge is zero, so is the field.

$$\frac{\partial \phi}{\partial r} = \frac{-q_{enc}}{2\pi \epsilon_s \epsilon_0 r^2} \tag{10.5}$$

where $\phi$ is potential, $r$ is the radial coordinate, $q_{enc}$ is the enclosed charge, and $\epsilon_s \epsilon_0$ is the absolute permittivity of the medium.

Since the potential decays roughly exponentially in a double layer, the double layer is only a few Debye lengths thick. If the double layer is very small on the relevant diffusional scales of the system, i.e. if the Debye length is much shorter than the size of the electrode, then the radius of the 'outside' of the double layer is coincident with the electrode radius to a good approximation.

Since the Debye length is of the order of tens of nanometres even at weak support, this is certainly the case for a microelectrode with $r_e \geq 1\ \mu m$. Therefore, weakly supported voltammetry can be simulated using the Poisson equation to describe the electric field, with the boundary condition $\partial \phi / \partial r = 0$ at the electrode surface. This is useful since we can simulate the problem without recourse to electroneutrality, but also without any detailed knowledge of the double layer structure.

(ii) The zero-field approximation is not appropriate for weakly supported voltammetry at nanoelectrodes since in this case it is no longer true that the Debye length is much smaller than the diffusion layer thickness. Therefore, the double layer and diffusion layer from the electrochemical reaction are likely to be convoluted in such a manner that the zero-field approximation can no longer be accurately applied at the boundary $r = r_e$.

## 10.9  Self-Supported Reduction of the Cobaltocenium Cation

### Problem

(i) How does the ohmic drop between an electrode and bulk solution affect the term for 'overpotential' that appears in both the Nernst and Butler–Volmer equations?

(ii) Using this, and employing the electroneutrality approximation that $\sum_i z_i c_i = 0$ for all species $i$ in a system, determine the *steady-state* ohmic drop as a function of overpotential for the reduction of a cobaltocenium salt (AX where $A^+$ is cobaltocenium and $X^-$ is its counter-ion) at a hemispherical electrode in the absence of any supporting electrolyte. Assume that all ions have equal diffusion coefficients. The relevant reaction is

$$A^+ + e^- \rightleftharpoons B^0$$

(iii) Hence determine the steady-state current at high overpotential, and comment on its magnitude.

## Solution

(i) The effect of ohmic drop is to resist the passage of current, and hence it reduces the overpotential 'perceived' by the working electrode. The relevant overpotential, $E_{obs}$, which is the driving force for a reaction, is the potential difference between the working electrode, $\phi_M$, and the solution at the plane of electron transfer, $\phi_{s,PET}$, measured versus a reference potential difference, $\Delta\phi_{ref}$:

$$E_{obs} = (\phi_M - \phi_{s,PET}) - \Delta\phi_{ref}$$

The potential difference applied, however, is the potential difference to bulk solution, measured versus a reference potential difference:

$$E = (\phi_M - \phi_{s,bulk}) - \Delta\phi_{ref}$$

So long as the solution is well supported, an alteration to $E$ produces an equivalent alteration to $(\phi_M - \phi_{s,bulk})$. If there is a significant potential difference between the plane of electron transfer and bulk solution, however, then:

$$\begin{aligned} E_{obs} &= (\phi_M - \phi_{s,PET}) - \Delta\phi_{ref} \\ &= E + (\phi_{s,bulk} - \phi_{s,PET}) \\ &= E - \Delta\phi_{OD} \end{aligned}$$

where $\Delta\phi_{OD}$ is the potential difference between the plane of electron transfer and bulk solution, i.e. the ohmic drop. Therefore, in all equations involving $E$, it should be substituted by $(E - \Delta\phi_{OD})$ to account for the resistance due to the ohmic drop.

(ii) We shall use the normalised concentration $c_i$ where the real concentration is normalised to bulk concentration. We also use the normalised potential $\theta = (F/RT)\phi$, and a radial coordinate $R = r/r_e$. Therefore, initially $c_A = c_X = 1$ and $c_B = 0$.

We need to solve a set of steady-state Nernst–Planck equations:

$$\frac{dc_A}{dR} + c_A\frac{d\theta}{dR} = \frac{1}{D_A}\frac{J_A}{R^2}$$

$$\frac{dc_B}{dR} = \frac{1}{D_B}\frac{J_B}{R^2}$$

$$\frac{dc_X}{dR} - c_X\frac{d\theta}{dR} = \frac{1}{D_X}\frac{J_X}{R^2}$$

where the fluxes $J_i$ have been defined *into* the electrode.

From the electroneutrality approximation

$$c_A - c_X = 0$$

and therefore $c_A = c_X$ everywhere.

At steady state the fluxes $J_i$ must be constant in $R$, and so since at $R = 1$ (the electrode surface) X is inert, $J_X = 0$ there and everywhere. Equally, from conservation of mass, $J_A = -J_B$ at the electrode and so the same is true everywhere.

Then if all diffusion coefficients are equal, $D_A = D_B = D_X = D$, we can sum the Nernst–Planck equations:

$$\frac{\partial(c_A + c_B + c_X)}{\partial R} + (c_A - c_X)\frac{\partial\theta}{\partial R} = \frac{1}{D}\left(\frac{J_A - J_A + 0}{R^2}\right)$$

therefore

$$\frac{\partial(c_A + c_B + c_X)}{\partial R} = 0$$

$$c_A + c_B + c_X = \gamma$$

where $\gamma$ is a constant. From bulk concentrations, $\gamma = 2$, and since $c_A = c_X$ everywhere:

$$2c_A + c_B = 2$$

$$c_B = 2(1 - c_A)$$

Also, from the Nernst–Planck equation in $c_X$ and the requirement $c_A = c_X$ it follows that

$$c_A = \exp(\theta)$$

as was shown above (Problem 10.6, part ii.).
The Nernst equation requires that at $R = 1$:

$$c_{A,0} = c_{B,0}\, e^{\theta - \theta_0}$$

where $\theta \equiv (F/RT)(E - E_f^{\ominus})$ and $\theta_0 = (F/RT)\,\Delta\phi_{OD}$.
Substituting

$$c_{A,0} = 2(1 - c_{A,0})\, e^{\theta - \theta_0}$$

$$e^{\theta_0} = 2(1 - e^{\theta_0})\, e^{\theta} e^{-\theta_0}$$

and rearranging

$$e^{2\theta_0} + 2e^{\theta} e^{\theta_0} - 2e^{\theta} = 0$$

We can let $x = e^{\theta_0}$ and so

$$x^2 + 2e^{\theta} x - 2e^{\theta} = 0$$

Since $x > 0$ by definition, since $\theta_0$ must be real:

$$x = e^{\theta_0} = \sqrt{e^{2\theta} + 2e^{\theta}} - e^{\theta}$$

Therefore,

$$\theta_0 = \ln\left(\sqrt{e^{2\theta} + 2e^{\theta}} - e^{\theta}\right)$$

(iii) We can also determine $c_B$ by integrating its Nernst–Planck equation which is linear and separable, since B is neutral:

$$c_B = \frac{J_A}{R}$$

therefore,

$$c_A = 1 - \frac{J_A}{2R}$$

and

$$c_{A,0} = e^{\theta_0} = 1 - \frac{J_A}{2}$$

Substituting the expression for $e^{\theta_0}$ above, and rearranging:

$$J_A = -2\left(\sqrt{e^{2\theta} + 2e^{\theta}} - e^{\theta} - 1\right)$$
$$= 2\left(1 + e^{\theta} - \sqrt{e^{2\theta} + 2e^{\theta}}\right)$$

As $\theta \to -\infty$, $e^{\theta} \to 0$ and so $J_A \to 2$, as compared to the limiting value of 1 at full support. This reflects attractive migration contributing to the steady-state current: the reduction of $A^+$ creates a layer of uncompensated negative charge close to the electrode, which attracts further $A^+$ and so elevates the mass transport-limited current.

# 11

## Voltammetry at the Nanoscale

### 11.1 Debye Length vs Diffusion Layer Thickness

### Problem

As was discussed in Chapter 5, diffusion towards a spherical electrode will attain a steady state at long times after a potential step. The expression for the concentration profile, following a potential step to a potential where electron transfer is rapid, is as follows:

$$c = c^* \left( 1 - \frac{r_e}{r} \operatorname{erfc}\left( \frac{r - r_e}{\sqrt{4Dt}} \right) \right)$$

where $c^*$ is the bulk concentration, $r_e$ is the electrode radius, and $D$ is the diffusion coefficient.

(i) Assess the typical size of the diffusion layer as time, $t$, tends to infinity, as a function of $r_e$.

(ii) The Debye length, the length over which electric fields are screened by a double layer, is given

$$r_D = \sqrt{\frac{RT\epsilon_s\epsilon_0}{2F^2 I}}$$

where $R$, $T$ and $F$ take their usual meanings, $\epsilon_s$ is the relative permittivity of the medium ($\approx 78.5$ for water at 25°C), $\epsilon_0$ is the vacuum permittivity ($= 8.854 \times 10^{-12}$ F m$^{-1}$) and $I$ is the ionic strength. Hence calculate Debye lengths for aqueous solutions of 0.1 M and 1 M binary electrolyte AX (such as a typical supporting electrolyte), at 298 K.

(iii) At what electrode sizes are the diffusion layer thickness and the Debye length similar in magnitude? Why is this significant for nanoscale electrochemistry?

## Solution

(i) Where $t$ is large, the expression within the erfc function tends to zero, and hence, since $erfc(0) = 1$:

$$\frac{c}{c^*} \rightarrow \left(1 - \frac{r_e}{r}\right)$$

Therefore, when $c/c^*$ attains some critical fraction $\alpha$, implying a certain degree of transition from the diffusion layer to bulk solution, $1 - (r_e/r) = \alpha$. Hence

$$r = \frac{r_e}{\left(1 - \frac{c}{c^*}\right)}$$

so for $\alpha = 0.9$, where bulk solution is almost attained, this point occurs at $r = 10r_e$. From this, we can determine that the size of the steady-state (long-time) diffusion layer scales proportionally to the electrode radius.

From the special case $\alpha = 0.5$, we note that where $c = (c^*/2)$, $r = 2r_e$. Hence, over the distance of one electrode radius away from the electrode surface, the concentration rises from total depletion to one half of the bulk value.

If the electrode has a radius of the order 10 nm, the diffusion layer is concentrated within a few tens of nanometres away from the electrode surface. This is in contrast to a microelectrode where the diffusion layer will be microns thick, or a macroelectrode where the steady-state diffusion layer is so large that the steady-state limit is not achieved on an experimental timescale. Instead, the diffusion layer continues to grow at a rate proportional to $\sqrt{Dt}$, throughout the course of an experiment.

(ii) For a binary electrolyte, the concentration of electrolyte and the ionic strength are equivalent. Hence:

$$r_D = \sqrt{\frac{RT\epsilon_s\epsilon_0}{2F^2}} \frac{1}{\sqrt{1000 \times c^*}}$$

where $c$ is given in M and the factor of 1000 converts to mol m$^{-3}$. Substituting in the physical constants provided:

$$r_D = 9.617 \times 10^{-9} \times \frac{1}{\sqrt{1000 \times c^*}}$$

Hence when $c^* = 0.1$, $r_D = 0.9617$ nm, and when $c^* = 1$, $r_D = 0.3041$ nm. For a well-supported aqueous solution, the Debye length is approximately one nanometre or less.

(iii) The Debye length becomes similar in magnitude to the steady-state diffusion layer thickness only for electrodes with dimensions of approximately 1 nm (unless the solution is weakly supported), but it may correspond to a significant fraction of the diffusion layer thickness for electrodes smaller than about 50 nm.

In these cases, the double layer and the diffusion layer are no longer straightforwardly decoupled, and so electric fields, altered overpotentials and population differences in the double layer are likely to affect the voltammetry in a manner which would not occur for electrochemistry at a larger electrode. Therefore, nanoscale electrochemistry requires consideration of the double layer and is theoretically more demanding to interpret, as well as opening the door to a number of interesting phenomena.

## 11.2 Altered Electrode Kinetics and Reactivity at the Nanoscale

### Problem

It has been observed for silver nanoparticles that $k^0$ and $\alpha$ values are altered at nanoelectrodes from their values for the bulk material, as is the case for 4-nitrophenol reduction in water, as compared to the reaction at bulk silver [F.W. Campbell *et al.*, *ChemPhysChem.* **11** (2010) 2820].

Further, the electrochemistry may be entirely changed, such as the altered mechanism of peroxide reduction in acidic media between nanoparticles and bulk material [F.W. Campbell *et al.*, *J. Phys. Chem. C* **113** (2009) 9053].

 (i) Explain why nanoparticles can exhibit such altered behaviour, especially when adsorption is involved in the electrochemical reaction.
(ii) What features of the voltammetry of nanoparticles supported on an unreactive substrate make the use of numerical simulation essential in the analysis of such voltammetry?

### Solution

 (i) Nanoparticles exist in constrained geometries due to their small size. As a consequence, a range of different crystal planes are exhibited at the surface, including those that may not be encountered in the bulk material. Additionally, no single surface plane will be particularly extensive due to the curvature of the surface of a nanoparticle, and therefore boundaries and defects are likely to be more common than for the bulk material. The exception to the above is the case of highly ordered nanoparticles which retain precise cubic or prismatic structures, despite their small size.

Where the surface structure of a nanoparticle differs from the bulk material, different adsorption kinetics are expected. This can be sufficient to alter the mechanism of a reaction where adsorption is significant, such as the reduction of peroxide in acidic media. The difference in solvation structure, such as an alteration of the compact portion of the double layer, and the possible difference in band structure of the electrode can lead to a change in the nature of the potential energy surface associated with electron transfer, and hence a change in the $k^0$ and $\alpha$ parameters of the Butler–Volmer equation.

(ii) While mass transport to an individual nanoparticle might be described using simple theories assuming a hemispherical nanoelectrode, it is rarely the case that an individual nanoparticle is studied – the current drawn would typically be too small compared to system noise. When nanoparticles are deposited onto a substrate by a method such as evaporation of a droplet of a nanoparticle suspension, a large number of nanoparticles will be distributed across the surface. The system therefore corresponds to a random array of nanoelectrodes, and hence numerical simulation is required to describe effects resulting from diffusion layer overlap; the diffusion domain approximation discussed in Chapter 6 is often invoked.

In fact, nanoparticles often cluster, rather than distributing in a perfectly random manner. An alteration to the distribution used in weighting different diffusion domains may be able to account for this. It is necessary to examine the electrode by a surface sensitive technique such as STM or AFM in order to accurately assess the distribution and coverage of nanoparticles, and hence to accurately describe the diffusional regime.

## 11.3 Nanoparticles: Case 4 Behaviour

### Problem

(i) How is the diffusional Case 4 characterised (see Chapter 6 of this text and of *Understanding Voltammetry*)? Why do many nanoparticle arrays exhibit diffusional Case 4 behaviour?

(ii) Figure 11.1 shows voltammograms recorded at $v = 1\,\mathrm{V\,s^{-1}}$ for a one-electron oxidation of 1 mM analyte at a) an array with a 1% coverage of 1 $\mu$m electroactive discs on a substrate, and b) for a 1% coverage of 10 nm nanoparticles on a substrate (approximated as having the same geometry, i.e. a regular array of discs). Discuss the differences in the voltammetry in the context of the relative particle sizes.

### Solution

(i) The diffusional Case 4 is the case in which active sites of electron transfer, such as nanoparticles, are separated by a very short distance compared to the distance over which a particle can diffuse during the experiment. Therefore,

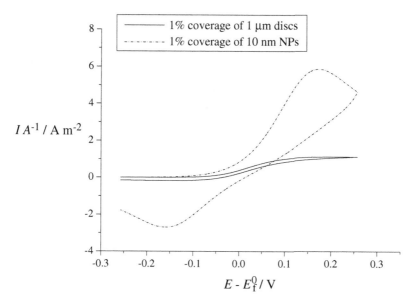

**Fig. 11.1** Simulated voltammetry for a one-electron oxidation of 1 mM analyte at two arrays, with current density $IA^{-1}$ given with reference to substrate area. $D$ is $10^{-5}$ cm$^2$ s$^{-1}$.

diffusion layers resulting from electrolysis overlap extensively, and an overall diffusion layer exists which is roughly planar, reflecting the geometry of the substrate rather than of any individual nanoparticle.

Because of the dominance of planar diffusion, voltammetry resembles that at a macroelectrode. Distinct forward and backward peaks are observed, and the peak current will scale with the square root of scan rate according to the Randles–Ševčík equation. However, the current reflects the geometric area of the substrate, not of the nanoparticles themselves, even though only a tiny fraction of this area is electroactive.

Many nanoparticle arrays exhibit diffusional Case 4 behaviour because the nanoparticles are deposited in sufficient density on the substrate that their diffusion layers overlap extensively. If the nanoparticles are being exploited for their catalytic properties, this is highly economical by comparison to the use of a macroelectrode of a catalytic metal, since only a very small quantity of the catalyst is required to achieve an equivalent current.

(ii) While the microdisc array shows Case 2/3 voltammetry (sigmoidal), the nanoparticle array shows Case 4 voltammetry with a distinct forward peak and marked peak-to-peak separation due to the relatively slow kinetics of the system. This can be understood since the small size of the nanoparticles means that individual sites of electron transfer are separated by much shorter distances than for the microdisc array, even though the proportional coverages are the same.

At $1\,V\,s^{-1}$, the forward sweep of the voltammogram lasts about 0.15 s. In this time, a mean diffusional distance is $\sqrt{2Dt}$ from the Einstein equation, which is about $17\,\mu m$. For a coverage of 1%, the diffusion domain radius is $10\,r_e$. Hence, the overlap of the diffusion layers is only partial in the case of $r_e = 1\,\mu m$, and so the voltammetry exhibits Case 3 diffusional behaviour, whereas for $r_e = 10\,nm$ the overlap is complete and Case 4 behaviour is observed.

Note that the current density is measured here in terms of the substrate area rather than the electroactive area, and hence is maximised in Case 4 when all available material is oxidised, rather than there existing zones between electrodes where electroactive material does not react as in Cases 2/3.

## 11.4 'Coulomb Staircase' Effects

### Problem

The charging of very small nanoparticles has been known to exhibit a 'Coulomb staircase' behaviour, where individual quantised electron transfers are observable.

Consider a double layer with a typical interfacial capacitance (per unit area) of $C_{DL} = 50\,\mu F\,cm^{-2}$.

(i) If this double layer describes the ideal capacitive charging at the surface of a hemispherical nanoelectrode which is driven to a potential of 25 mV vs its potential of zero charge, what size of electrode corresponds to a charge of a) 100 electrons and b) 10 electrons?

(ii) Hence describe the likely appearance of the capacitive charging transient as the potential is swept away from the potential of zero charge at a 5 nm radius electrode, with the interfacial capacitance properties given above.

### Solution

(i) An ideal capacitor obeys the relation:

$$C = \frac{q}{E}$$

where $C$ is capacitance, $q$ is charge separation and $E$ is applied potential. Therefore,

$$q = CE$$
$$= 50 \times 10^{-6} \times 10^4 \times 2\pi \times r_e^2 \times 0.025\,C$$
$$= 0.07854\,r_e^2\,C$$

where $2\pi r_e^2$ gives the area of the double layer as the surface area of the hemispherical nanoelectrode, and the factor of $10^4$ converts $C_{DL}$ into metre units. Now, dividing by the charge on an electron and rearranging

$$n_{e^-} = \frac{0.07854\, r_e^2}{1.602 \times 10^{-19}}$$

$$r_e = \sqrt{\frac{n_{e^-} \times 1.602 \times 10^{-19}}{0.07854}}$$

$$= \sqrt{n_{e^-}} \times 1.428 \,\text{nm}$$

For 100 electrons, the required size is 14.28 nm; for 10 electrons, the required size is 4.516 nm.

(ii) For a 5 nm radius electrode that charges ideally at low voltage and has a capacitance given as $50\ \mu F\ cm^{-2}$, approximately 10 electrons are added as excess charge per 25 mV separation from the potential of zero charge. The spacing between discrete (quantised) electron transfers is then approximately 2.5 mV. In place of the familiar capacitive current ramp, current peaks are expected with a 2.5 mV spacing; on integration, discrete steps of increasing electrode charge may be discernable.

## 11.5 Ultrafast 'Single Molecule' Voltammetry

## Problem

In a series of voltammetric experiments on nanoscale dendrimers, C. Amatore *et al.* discussed interesting nanoscale effects in single molecule voltammetry [*ChemPhysChem* **2** (2001) 130]. Fourth-generation PAMAM (polyamidoamide) dendrimers modified with 64 Ru(II) redox sites at their surface were adsorbed onto a platinum substrate: the dendrimer molecule has a radius of approximately 5 nm (see Fig. 11.2).

(i) To study very short length scales, megavolt-per-second voltammetry was employed. Why is ohmic drop particularly problematic when the scan rate is extremely fast?

(ii) Give an expression for the diffusion layer thickness as a function of scan rate, $v$, considering some region of the potential sweep of length $\Delta E$ over which the diffusion layer is growing.

(iii) If $\Delta E \approx 100$ mV and electron hopping between redox sites on the dendrimer surface exhibits an apparent diffusion coefficient of $D = 5 \times 10^{-6}$ cm$^2$ s$^{-1}$, why might megavolt-per-second scan rates cause voltammetric behaviour which differs from that of an ideally adsorbed redox species?

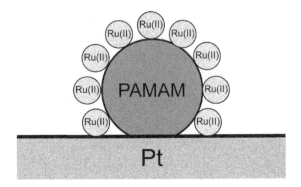

**Fig. 11.2** Simplified schematic of a Ru(II)-bisterpyridyl modified fourth-generation PAMAM dendrimer adsorbed onto a Pt substrate. 'Ru(II)' indicates the Ru-containing functional group.

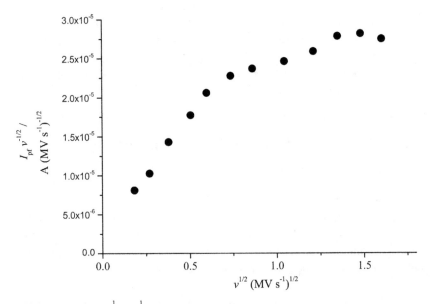

**Fig. 11.3** Plot of $I_{pf} v^{-\frac{1}{2}}$ vs $v^{\frac{1}{2}}$ for fast voltammetry at the dendrimer-modified Pt electrode, indicating a change in diffusional regime. Note that the plot is near linear at low $v$, indicating $I_{pf} \propto v$. Adapted from C. Amatore *et al.*, *ChemPhysChem* **2** (2001) 130, with permission from Wiley.

(iv) Positive feedback ohmic drop compensation was used to record voltammetry for the system. Peak currents are shown in Fig. 11.3. Discuss the trend(s) exhibited.

## Solution

(i) For both the voltammetry of adsorbed and diffusing species, current increases with scan rate, since the consumption of the immediately available material at the electrode surface is driven over an increasingly short time.

According to Ohm's law, the potential difference arising from driving a current $I$ through a medium with resistance $R_s$ is $IR_s$. Hence the overpotential associated with overpotential increases with driven current, and so with scan rate. At megavolt-per-second scan rates, specialised equipment is required to disentangle useful information given by Faradaic currents from current distortions due to resistive and capacitive effects.

(ii) From the Einstein equation, we can describe the mean diffusion layer thickness as

$$x_{\text{diff}} = \sqrt{2Dt}$$

where

$$t = \frac{\Delta E}{v}$$

Therefore

$$x_{\text{diff}} = \sqrt{\frac{2D\Delta E}{v}}$$

(iii) Substituting into the expression above:

$$x_{\text{diff}} = \sqrt{\frac{2 \times 5 \times 10^{-10} \times 0.1}{v}}$$

$$= \sqrt{\frac{10^{-10}}{v}}\ \text{m}$$

Hence if $v = 1\ \text{MV s}^{-1}$, it follows that

$$x_{\text{diff}} = \sqrt{\frac{2 \times 10^{-10}}{2 \times 10^{6}}} = 10^{-8}\ \text{m} = 10\ \text{nm}$$

The mean diffusion layer thickness in this case is almost exactly the diameter of the dendrimer molecule. It therefore follows that the oxidation of all redox sites will not be complete because a 'diffusion layer' of oxidation achieved by successive electron hopping across the dendrimer surface is not yet complete.

At faster scan rates this effect will be more extreme; at slower scan rates, in the kilovolt-per-second range, the 'diffusion layer' associated with this $D$ for electron hopping is sufficiently large that all redox sites are equilibrated with the potential applied to the substrate, and so the voltammetry appears as normal for an adsorbed species.

(iv) This plot shows the transition from ideal adsorbed species voltammetry, where $I_{pf} \propto v$, to diffusional voltammetry where the 'diffusion process' is electron hopping between redox sites on the surface of the dendrimer. The latter regime is characterised by $I_{pf} \propto v^{\frac{1}{2}}$, as discussed in Chapter 4, and this is observed in the data. Note that in the original research, these data were used together with a more rigorous model of the 'electron diffusion' to derive the value of $D = 5 \times 10^{-6}$ cm$^2$ s$^{-1}$ for electron tunnelling quoted above.

## 11.6 Thin-Layer Effects in Nanoscale Voltammetry

### Problem

It has been shown [M.C. Henstridge *et al.*, *Sensor. Actuat.* B **145** (2010) 417] that the effect of a porous thin layer on cyclic voltammetry can be easily confused with electrocatalytic behaviour, since both effects cause a shift of peak potential to lower overpotentials. Such a thin layer may exist in, for instance, a multi-walled carbon nanotube (MWCNT)-modified electrode where the MWCNTs form a dense mesh across the substrate, a system which is also often associated with electrocatalysis.

 (i) Why is the voltammetry of an electroactive species dissolved in a thin layer (compared to mean diffusion distance during the scan) adjacent to the electrode very similar to that for an adsorbed species?
 (ii) Hence, explain why the peak potential will occur at lower overpotential than for diffusional voltammetry in such a system.
 (iii) How does the MWCNT-modified electrode behave like a thin layer?
 (iv) How might it be possible to identify whether the effect causing a peak potential shift arises from kinetics or diffusion?

### Solution

 (i) Both a thin layer of solution and an adsorbed layer have in common that only a finite quantity of material is available for electrolysis reaction. What is more, in both cases there is effectively no delay in replenishing material at the surface. Whereas in the adsorbed case the material is strictly confined to the electrode surface, a porous thin layer implies that diffusion across the layer is extremely fast compared to the timescale of the experiment. Therefore, the concentration is quickly equilibrated such that it is uniform through the layer.

 So, diffusion does not contribute to the observed voltammetry; the forward peak in the voltammetry arises due to the depletion of available material for reaction and is ideally at $E = E_f^{\ominus}$ if the electrode kinetics are fast, as was discussed in Chapter 6.

(ii) In diffusional voltammetry with an ample supply of bulk solution, the peak potential arises because of depletion of material at the electrode arising as the rate of reaction at the electrode surface exceeds the rate of incident diffusion. However, this diffusion allows the current to increase until past the formal potential of the reaction, since a certain quantity of material is able to be replenished in the course of the voltammetric wave.

This is not so for thin-layer voltammetry – as with the voltammetry of an adsorbed species, the finite quantity of solution and the rapid equilibration of concentration in that layer means that there is no significant replenishment of the surface concentration of the electroactive species. As such, rate-limiting depletion becomes significant more quickly and the peak potential will occur at lower overpotential.

(iii) The MWCNT-modified electrode resembles a thin layer because the deposited nanotubes form a dense but porous mesh on the substrate surface. The solution will permeate this mesh, and therefore a much larger quantity of the dissolved electroactive species begins the experiment in close proximity to an electroactive surface than is the case for a non-porous geometry.

As the overpotential becomes sufficient to drive a reaction, this material is quickly consumed and the concentration becomes uniform between the closely neighbouring nanotubes, as in a thin layer. Once the material within the thin layer is exhausted, reaction can only continue by diffusion from bulk, which is much slower. Therefore, the current reaches a peak at a lower overpotential than would be the case if the electroactive material were non-porous. This is demonstrated by simulated voltammetry at Fig. 11.4.

(iv) Note that the shifting of a peak to lower overpotential is also associated with increased $k^0$, i.e. electrocatalysis, a property which certainly is exhibited by many nanomaterials. The confusion of these effects is particularly likely if experiments are conducted only at a single scan rate.

Adsorption or thin-layer processes can be characterised by a peak current dependence on scan rate which is linear: $I_{pf} \propto v$. Diffusional processes, by contrast, are characterised in the Case 4 limit by a square-root dependence: $I_{pf} \propto v^{\frac{1}{2}}$. By comparison of the scan rate dependence of experimental peak current to these equations, across a wide range of scan rates, it is possible to ascertain if the voltammetry is diffusional or depends on surface effects such as adsorption or the formation of a thin layer.

Unfortunately, the role of adsorption is not straightforwardly disentangled. This is especially true if the adsorption is reversible. If not, the presence of an adsorbed layer might be determined by transferring the electrode to a clean solution and repeating the experiment, as was discussed in M.C. Henstridge *et al.* (see above).

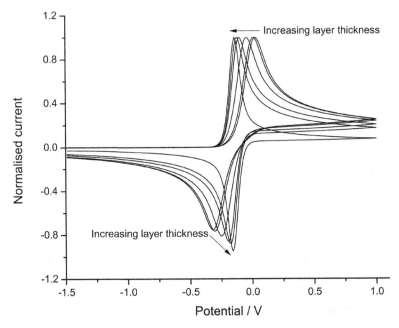

**Fig. 11.4** Apparent 'electrocatalytic' effect on forward peak potential as thin layer diffusion begins to dominate conventional diffusion. Voltammetry has been normalised to peak current for clarity. Reproduced from M.C. Henstridge *et al.*, *Sens. Actuators B* **145** (2010) 417, with permission from Elsevier.

## 11.7 Voltammetry in a Nanochannel

### Problem

M.A.G. Zevenbergen *et al.* have reported voltammetry performed in a 50 nm high nanochannel [*J. Am. Chem. Soc.* **131** (2009) 11471]. At both top and bottom of the nanochannel are Pt electrodes, which are planar and extend a much larger distance than the height of the channel in the other two dimensions. The applied potential may be set separately at each electrode.

(i) On the assumption that the device rapidly attains steady state, as should be expected given its size, what is the maximum flux for the oxidation of 1 mM Fc(MeOH)$_2$ that may be drawn with one electrode at a highly oxidising potential and one at a highly reducing potential? You may take the diffusion coefficient of Fc(MeOH)$_2$ as $6 \times 10^{-6}$ cm$^2$ s$^{-1}$ and you may assume that the solution is well supported (only diffusional transport).

(ii) Hence calculate the incident velocity of material towards the working electrode, and suggest the largest $k^0$ that can be resolved.

(iii) From this velocity, determine the mean time required for a diffusing molecule to traverse the channel from top to bottom. At a scan rate of $10 \, \text{mV s}^{-1}$, how many times might such a molecule move back and forth in the course of a scan across $100 \, \text{mV}$?

(iv) Under various conditions, 'redox cycling' was shown to markedly elevate the current. Discuss what is meant by redox cycling. Why is redox cycling dependent on a potential difference being applied vs a reference at *both* electrodes? What factors may limit the current drawn by redox cycling?

## Solution

(i) The geometry of the nanochannel allows us to neglect transport in all directions other than perpendicular to the electrode, which we denote $x$. At steady state, where $\partial c / \partial t = 0$, the following equation describes the concentration profile according to Fick's second law:

$$\frac{d^2 c}{dx^2} = 0$$

Hence

$$c = Ax + B$$

If the electrode at $x = 0$ oxidises $\text{Fc(MeOH)}_2$ rapidly, $c = 0$ here. Equivalently, if the electrode at $x = \Delta x$ reduces $\text{Fc(MeOH)}_2^+$ rapidly, $c = c^*$ here. Therefore:

$$c = \frac{c^*}{\Delta x} x$$

From Fick's first law

$$J = -D\frac{dc}{dx} = -\frac{Dc^*}{\Delta x}$$

where the negative value indicates flux towards the electrode at $x = 0$. Substituting in the known values for these parameters, and converting to metre units in all cases

$$|J_{\lim}| = \frac{6 \times 10^{-10} \times 1}{5 \times 10^{-8}} = 0.012 \, \text{mol m}^{-2} \, \text{s}^{-1}$$

(ii) Note that the limiting flux is dependent on the concentration of the solution. We can determine a critical velocity of transport towards the surface as:

$$v = \frac{J}{c^*} = 0.012 \text{m s}^{-1}$$

Therefore the diffusional transport in the nanochannel compares to typical 'fast' $k^0$ values of $1 \, \text{cm s}^{-1} = 0.01 \, \text{m s}^{-1}$. In fact, somewhat faster kinetics can still be observed, since the diffusional rate remains within an order of

magnitude of the kinetic rate. The experimental results of M.A.G. Zevenbergen *et al.* indicated clear and reproducible experimental distinction of $k^0$
values up to 15 cm s$^{-1}$ for the Fc(MeOH)$_2$ system.

(iii) We can crudely define

$$t_{\text{diff}} = \frac{\Delta x}{v}$$

$$= \frac{5 \times 10^{-8}}{0.012}$$

$$\approx 4\,\mu\text{s}$$

One can then imagine a single molecule undertaking a series of 8 $\mu$s back-
and-forth trips between the electrodes, of which a little over one million are
hence possible in the 10 s required to sweep 100 mV at the given rate.

(iv) The calculation above gives some idea of how such large fluxes can be achieved
within a very small volume of solution. If the top electrode is disconnected,
the current is limited because the nanochannel then behaves as a thin layer
and the material available for oxidation is rapidly exhausted. The current is
then rate-limited by the rate of diffusion into the channel, which is very small.

If, however, reduction of the product species is driven by applying a potential at the top electrode, vs an independent reference electrode, the material
available for oxidation at the bottom electrode is replenished.

'Redox cycling' describes such a cycle for a single molecule, which is oxidised, diffuses a short (nano-) distance, is reduced, and diffuses back to be
oxidised again (see Fig. 11.5). By achieving multiple electron transfers per
molecule, currents can be large even with a limited volume of solution and a
limited electrode area. F.-R.F. Fan *et al.* achieved this effect by using a sharp tip
with an insulating sheath to trap single molecules in a nanovolume between
the oxidising tip and a reducing substrate [*Science* **267** (1995) 871].

The advantage of the nanochannel method is that due to the relatively
large size of the channel in the other two directions, a large number of
molecules may simultaneously undergo redox cycling, yielding relatively easily

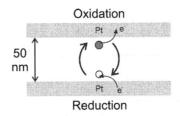

**Fig. 11.5** Schematic of redox cycling in a 50 nm nanochannel.

measurable nanoamp currents, rather than picoamp currents as with the single molecule studies. Note that molecules are still undergoing lateral diffusion, however, so that some molecules may leave the active volume of the channel and cease to undergo redox cycling. This loss will limit the efficiency of the redox cycling process.

# Index